Salkind's book is in a class by itself. It is easily the best book of its kind that I have come across. I enthusiastically recommend it for anyone interested in the subject—and even (and especially) for those who aren't!

—Professor Russ Shafer-Landau
University of Wisconsin

I just wanted to send a little "thank you" your way for writing an extremely user-friendly book, *Statistics for People Who (Think They) Hate Statistics*. I'm a psychology major doing an independent study over break (at Alverno College, a statistics course is a prerequisite for a class I'm taking this spring, Experimental Psychology). In other words, I'm pretty much learning this on my own (with a little guidance from my mentor), so I appreciate having a book that presents the material in a simple, sometimes humorous manner.

—Jenny Saucerman

Dr. Salkind's book is a "must read" for students who think they don't "do statistics." He writes clearly about statistical topics and has a unique way of making them fun. The book contains useful explanations, examples that help students understand the underlying concepts, and graphics that clarify the material without overwhelming the novice.

—Professor Nancy Leffert
Fielding Graduate Institute

I just wanted to let you know how much I enjoyed your book *Statistics for People Who (Think They) Hate Statistics* and how easily it jogged my faint memories of statistics (my work had been primarily in the clinical world before returning to the academic world) and of SPSS. I am sure you hear from students all the time, but I wanted to let you know that even other academics find your book useful.

—Professor John T. Wu, EdD
Point Loma Nazarene University

Great presentations for a subject that tends to be esoteric—the text makes statistics alive and vibrant. I told my wife that the book reads like a novel—I can hardly put it down.

—Professor Kenrick C. Bourne
Loma Linda University

I love the clear description of the two-tailed test.

—Pepper
The author's dog

My students really appreciate your approach, which is making my job a lot easier.

—Professor Tony Hickey
Western Carolina University

I love your book *Statistics for People Who (Think They) Hate Statistics*. I thought I did hate statistics; to be honest, I feared the concepts of stats, numbers, math, etc. . . . Ewwww! But thanks to your book, I understand it now (I get it). Your book gives me hope. I'm working on my PhD in Nursing here in Baton Rouge, Louisiana, and I'm confident that I will ace my 100% online, graduate stats class. This text is my professor and guidance during these late nights of studying (my best time to study). This book is loaded with helpful tips and clarity, and it's fun. I love the part about the 100 airline pilots and the flying proficiency test. The lowest value was 60—"don't fly with this guy." Love it— funny. Thanks, Dr. Salkind.

—Del Mars

I studied statistics 20 years ago and recently moved from administration into health research. Your book has been a big help in reviewing basic statistics. I love the book! Please write another.

—Susan Lepre, PhD
Bergen County Department of Health Services

Hello! I bought your book at Barnes and Noble among 30 books that I browsed for my statistics class. I was intrigued by the title . . . and it was so simple to understand with the step ladder format. I followed those steps, and boy, they really work! Thanks a lot!

—Anne Marie Puentespina, RN, BSN
Legal Nurse Consultant

For my beginning students, this is the book that fits their needs. It is clear, concise, and not intimidating. It's even fun. I strongly recommend it.

—Professor Lew Marglois
School of Public Health, University of North Carolina

I have loved statistics ever since my second undergraduate course. Your book *Statistics for People Who (Think They) Hate Statistics* has cleared up confusion and partial understandings that I have had for years. It is a must for anyone beginning or continuing their journey in this science. I love it and will use it for all of the foreseeable future.

—Ronald A. Straube
Performance Improvement Coordinator,
Mission Texas Regional Medical Center

I am a doctoral student and simply love your book.

—Marisol Miller

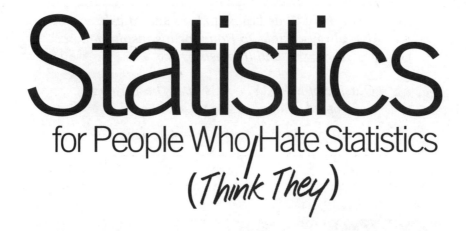

Statistics
for People Who Hate Statistics
(Think They)

SECOND EDITION

EXCEL 2007 EDITION

This book is dedicated to Sara and Micah.
Also, always thanks to Doug for his inspiration
and friendship.

Outside of a dog, a book is man's best friend.
Inside of a dog, it's too dark to read.

—Groucho Marx

"Two Tails Up"
In Memory of Pepper
1994—2009

Statistics

for People Who Hate Statistics

(*Think They*)

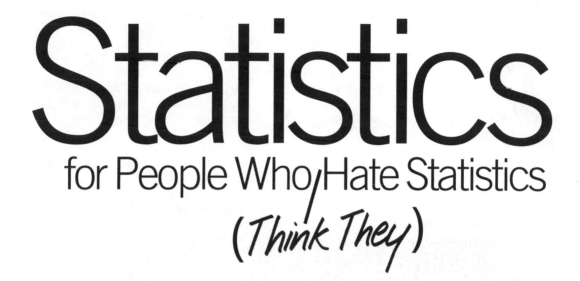

SECOND EDITION

EXCEL 2007 EDITION

Neil J. Salkind

University of Kansas

⑤SAGE

Los Angeles • London • New Delhi • Singapore • Washington DC

For information:

SAGE Publications, Inc.
2455 Teller Road
Thousand Oaks, California 91320
E-mail: order@sagepub.com

SAGE Publications Ltd.
1 Oliver's Yard
55 City Road
London EC1Y 1SP
United Kingdom

SAGE Publications India Pvt. Ltd.
B 1/I 1 Mohan Cooperative Industrial Area
Mathura Road, New Delhi 110 044
India

SAGE Publications Asia-Pacific Pte. Ltd.
33 Pekin Street #02-01
Far East Square
Singapore 048763

Printed in the United States of America.

Library of Congress Cataloging-in-Publication Data

Salkind, Neil J.
Statistics for people who (think they) hate statistics: Excel 2007 edition/Neil J. Salkind. — 2nd ed.
    p. cm.
This edition shows the students how to install the Excel Analysis ToolPak option (free) to earn access to a host of new and very useful analytical techniques.
Includes index.
ISBN 978-1-4129-7102-7 (pbk. : acid-free paper)
    1. Statistics. 2. Microsoft Excel (Computer file) I. Title.

HA29.S2365 2009
519.5—dc22   2008049630

Printed on acid-free paper

09 10 11 12 13   10 9 8 7 6 5 4 3 2 1

Acquiring Editor:	Vicki Knight
Associate Editor:	Sean Connelly
Editorial Assistant:	Lauren Habib
Production Editor:	Sarah K. Quesenberry
Copy Editor:	Liann Lech
Proofreader:	Wendy Jo Dymond
Indexer:	Molly Hall
Typesetter:	C&M Digitals (P) Ltd.
Cover Designer:	Ravi Balasuriya
Marketing Manager:	Stephanie Adams

BRIEF CONTENTS

PART V

DETAILED CONTENTS

PART II

PART III

PART IV

PART V

A NOTE TO THE STUDENT: WHY I WROTE THIS BOOK

It's always fun to take on a new project. The new version of Excel made revising the first edition of *Statistics for People Who (Think They) Hate Statistics: The Excel Edition* interesting and productive. I hope that I did justice to the program and justice to the students who I continue to enjoy teaching.

What many of them (*still* after all these years) have in common (at least at the beginning of the course) is a relatively high level of anxiety, the origin of which is, more often than not, what they've *heard* from their fellow students. Often, a small part of what they have heard is true—learning statistics takes an investment of time and effort (and there's the occasional monster for a teacher).

But most of what they've heard (and where most of the anxiety comes from)—that statistics is unbearably difficult and confusing—is just not true. Thousands of fear-struck students have succeeded where they thought they would fail. They did it by taking one thing at a time, pacing themselves, seeing illustrations of basic principles as they are applied to real-life settings, and even having some fun along the way. That's what I tried to do in writing all the editions of *Statistics for People Who (Think They) Hate Statistics,* and I tried even harder in completing this revision.

After a great deal of trial and error, and some successful and many unsuccessful attempts, I have learned to teach statistics in a way that I (and many of my students) think is unintimidating and informative. I have tried my absolute best to incorporate all of that experience into this book.

What you will learn from this *Statistics for People . . .* is the information you need to understand what the field and study of basic statistics is all about. You'll learn about the fundamental ideas and the most commonly used techniques to organize and make sense out of data. There's very little theory (but some), and there are few mathematical proofs or discussion of the rationale for certain mathematical routines.

Why isn't this theory stuff and more in *Statistics for People Who (Think They) Hate Statistics?* Simple. Right now, you don't need it. It's not that I don't think it is important. Rather, at this point and time in your studies, I want to offer you material at a level I think you can understand and learn with some reasonable amount of effort, while at the same time not be scared off from taking additional courses in the future. I (and your professor) want you to succeed.

So, if you are looking for a detailed unraveling of the derivation of the analysis of variance F ratio, go find another good book from SAGE Publications (I'll be glad to refer you to one). But if you want to learn why and how statistics can work for you, you're in the right place. This book will help you understand the material you read in journal articles, explain what the results of many statistical analyses mean, and teach you how to perform basic statistical work.

And, if you want to talk about any aspect of teaching or learning statistics, feel free to contact me. You can do this through my e-mail address at school (njs@ku.edu). You can also keep up on anything new regarding this edition (and all versions of *Statistics for People . . .*) by going to www.statisticsforpeople.com.

Good luck, and let me know how I can improve this book to even better meet the needs of the beginning statistics student.

ACKNOWLEDGMENTS

Everybody, and I mean everybody (including Steve in shipping, Sarah Quesenberry and Liann Lech in production, and Stephanie Adams in marketing), at SAGE deserves a great deal of thanks for providing me with the support, guidance, and professionalism that takes only an idea (way back before the first edition) and makes it into a book like the one you are now reading, and then makes it successful.

However, there are some people who have to be thanked individually for their special care and hard work. C. Deborah Laughton supported the original idea for this type of book and Lisa Cuevas-Shaw for this particular book using Excel as a framework for teaching introductory statistics. Now, Vicki Knight is the executive editor who saw to it that this revision reached fruition, and she has provided the support and patience necessary to make this a reality. I am greatly appreciative. Associate Editor Sean Connelly's help has been invaluable as well. I also want to thank the following people (and no last names, but they know who they are) for their help in providing feedback on the previous edition as well as this edition: Dr. Rich Snow (OK, so now you know his last name) and his students, Gerrie D., Chris K., Bill B., Susan I., Dell M., Alice W., Katherine L. W., Dawn H., Len C., Rick F., Debbie S., Robin M.,

Lana B., Rick J., Andrew F., Brenda C., Gohar P., Kevin L., Thomas H., Gary A., John S., Ali A., Susana B., Nathaniel M., David R., Anne P., Frances L., Sara A., Nathaniel B., Kevin X., Cheryl K., and Joe D. Apologies to those I may have missed.

Also, special thanks go to William I. Bauer, Case Western Reserve University; Melanie E. L. Bush, Adelphi University; Lauri Hyers, West Chester University of Pennsylvania; Roger D. Lee, Salt Lake Community College; Jennifer Martin, Oakland University; Rob Mowrer, Angelo State University; Kevin M. Sowinski, Purdue University; and Minjuan Wang, San Diego State University who reviewed this edition. And, to Dan Ferguson who always helps with the online data sets.

AND NOW, ABOUT THE 2ND EDITION . . .

What you read above about this book reflects my thoughts about why I wrote this book in the first place. But it tells you little about this new edition.

Any book is always a work in progress, and the Excel edition of *Statistics for People Who (Think They) Hate Statistics* is no exception. Over the past 3 years or so, many people have told me how helpful this book is, and others have told me how they would like it to change and why. In revising this book, I am trying to meet the needs of all audiences. Some things remain the same, and some have changed.

There are always new things worth consideration and different ways to present old themes and ideas. Here's a list of what you'll find that's new in the 2nd edition of *Statistics for People Who (Think They) Hate Statistics: The Excel Edition.*

- There is a bunch of new exercises at the end of each chapter. Not only more, but also ones that vary greater in their level of application and (I hope) interest. As in earlier editions, these exercises use data sets that are available at www.sagepub.com/salkind2datasets and http://soe.ku.edu/Salkind/statsexcel_fp whs2e (from my school Web site). You can download them as needed. And, if you have any trouble downloading from either, or both, of these, e-mail me and I will send them to you immediately.
- The data sets are saved in two formats: Excel (from the 2007 Office suite) and what the Excel people call Excel 97-2003 Workbook, which works with earlier versions of Excel. Note

that this book focuses on the 2007 version of Excel and does not offer any instruction using earlier versions (that was for the 1st Excel edition of *Statistics for People . . .*).

- The answers to the Time to Practice questions continue to be at the end of the chapter. Some students and faculty prefer them to be at the end of the book in a separate appendix. Let me know what you find most useful.

- The information on reliability and validity has been moved to the first part of the book, rather than appearing later in the section on the part on learning about inferential statistics. This was done as a result of suggestions by several users that the material would be better understood if placed earlier in the book. This is the only major change in organization of the material.

- The 2nd edition features the latest version of Excel, from Office 2007. There are some pretty significant changes, so students (and others) might want to look through Appendix A, which is a quick guide to some main features. And, the Mac and Windows versions of Excel appear to be almost identical or very easy to use interchangeably, so users of this book can work on either platform except for one huge caveat! The Mac version does not offer the handy dandy Analysis Toolpak (discussed throughout the book). Yikes—what was Microsoft thinking? Instead functions and formulas can be used (but without having as much fun). Data files between the two are interchangeable as well.

Whatever typos and such appeared in any edition of this book are entirely my fault, and I apologize to the professors and students who were inconvenienced by their appearance. You can find a list of typos that will be corrected in a later printing at www.statistics forpeople.com. And I so appreciate all the letters, calls, and e-mails pointing out these errors and making this 2nd edition a better book for it. We have all made every effort in this edition to correct those and hope we did a reasonably good job. Let me hear from you with suggestions, criticisms, nice notes, and so on. Good luck.

Neil J. Salkind
University of Kansas
njs@ku.edu

PART I

Yippee!
I'm in Statistics

N ot much to shout about, you might say? Let me take a minute and show you how some very accomplished scientists use this widely used set of tools we call statistics.

- Michelle Lampl is a pediatrician and an anthropologist at Emory University. She was having coffee with a friend, who commented on how quickly her young infant was growing. In fact, the new mother spoke as if her son were "growing like a weed." Being a curious scientist (as all scientists should be), Dr. Lampl thought she might actually examine how rapid this child's growth, and others, is during infancy. She proceeded to measure a group of children's growth on a daily basis and found, much to her surprise, that some infants grew as much as one inch overnight! Some growth spurt.

  Want to know more? Why not read the original work? You can find more about this in Lampl, M., Veldhuis, J. D., & Johnson, M. L. (1992). Saltation and stasis: A model of human growth. *Science, 258,* 801–803.

- Sue Kemper is a professor of psychology at the University of Kansas and has been working on the most interesting of projects. She and several other researchers are studying a group of nuns and examining how their early experiences, activities, personality characteristics, and other information relate to their health during their late adult years. Most notably, this diverse group of scientists (including psychologists, linguists, neurologists, and others) wants to know how well all this information predicts the occurrence of Alzheimer's disease.

She's found that the complexity of the nuns' writing during their early 20s is related to the nuns' risk for Alzheimer's 50, 60, and 70 years later.

Want to know more? Why not read the original work? You can find more about this in Snowdon, D. A., Kemper, S. J., Mortimer, J. A., Greiner, L. H., Wekstein, D. R., & Markesbery, W. R. (1996). Linguistic ability in early life and cognitive function and Alzheimer's disease in late life: Findings from the nun study. *Journal of the American Medical Association, 275,* 528–532.

- Aletha Huston is a researcher and teacher at the University of Texas in Austin and devotes a good deal of her professional work to understanding what effects television watching has on young children's psychological development. Among other things, she and her late husband John C. Wright specifically investigated the impact that the amount of educational television programs watched during the early preschool years might have on outcomes in the later school years. They found convincing evidence that children who watch educational programs such as *Mr. Rogers* and *Sesame Street* do better in school than those who do not.

  Want to know more? Why not read the original work? You can find more about this in Collins, P. A., Wright, J. C., Anderson, R., Huston, A. C., Schmitt, K., & McElroy, E. (1997, April). *Effects of early childhood media use on adolescent achievement.* Paper presented at the biennial meeting of the Society for Research in Child Development, Washington, D.C.

All of these researchers had a specific question they found interesting and used their intuition, curiosity, and excellent training to answer it. As part of their investigations, they used this set of tools we call statistics to make sense out of all the information they collected. Without these tools, all this information would have been just a collection of unrelated outcomes. The outcomes would be nothing that Lampl could have used to reach a conclusion about children's growth, or Kemper could have used to better understand Alzheimer's disease, or Huston could have used to better understand the impact of watching television on young children's achievement and social development.

Statistics—the science of organizing and analyzing information to make the information more easily understood—made the task doable. The reason that any of the results from such studies are useful is that we can use statistics to make sense out of them. And that's exactly the goal of this book—to provide you with an understanding of these basic tools and how they are used—and, of course, how to use them.

In this first part of *Statistics for People Who (Think They) Hate Statistics . . . Excel 2007 Edition,* you will be introduced to what the study of statistics is about and why it's well worth your efforts to master the basics—the important terminology and ideas that are central to the field. It's all in preparation for the rest of the book.

We'll also be getting right into the Excel material with two little chapters that follow Chapter 1—one chapter about formulas and functions (Little Chapter 1a) and one chapter about the use of the Analysis ToolPak (Little Chapter 1b).

And this 2nd edition uses Excel 2007. Although you can use other versions with the material in the book and do OK, you'll be much better equipped to follow the material here and do the exercises if you have the current version of this pretty cool application.

1

Statistics or Sadistics?

It's Up to You

Difficulty Scale ☺☺☺☺☺ (really easy)

What you'll learn about in this chapter

- What statistics is all about
- Why you should take statistics
- How to succeed in this course

WHY STATISTICS?

You've heard it all before, right? "Statistics is difficult," "The math involved is impossible," "I don't know how to use a computer," "What do I need this stuff for?" "What do I do next?" and the famous cry of the introductory statistics student, "I don't get it!" Well, relax. Students who study introductory statistics find themselves, at one time or another, thinking about at least one of the above, if not actually sharing it with another student, their spouse, a colleague, or a friend.

And all kidding aside, there are some statistics courses that can easily be described as sadistics. That's because the books are repetitiously boring, and the authors have no imagination.

That's not the case for you. The fact that you or your instructor has selected *Statistics for People Who (Think They) Hate Statistics . . . Excel 2007 Edition* shows that you're ready to take the right approach—one that is unintimidating, informative, and applied (and even a little fun), and that tries to teach you what you need to know about using statistics as the valuable tool that it is.

4

If you're using this book in a class, it also means that your instructor is clearly on your side—he or she knows that statistics can be intimidating but has taken steps to see that it is not intimidating for you. As a matter of fact, we'll bet there's a good chance (as hard as it may be to believe) that you'll be enjoying this class in just a few short weeks.

And Why Excel?

Simple. It's the most popular, most powerful spreadsheet tool available today, and it can be an exceedingly important and valuable tool in learning how to use statistics. In fact, many stat courses taught at the introductory level use Excel as their primary computational tool and ignore other computer programs, such as SPSS and MiniTab. Although we are not going to teach you how to use Excel (see Appendix A for a refresher on some basic tasks), we show you how to use it to make your statistics learning experience a better one.

But like any program that takes numbers and consolidates and analyzes them, Excel is not a magic bullet or a tool to solve all your problems. It too has its limitations. Unless you are an expert programmer and you can program Excel to do just about anything other stat programs can (and the language you would use is called Visual Basic), Excel may not look as pretty as other programs dedicated to statistical analysis or offer as many of the same options. But at the level of introductory statistics, it is a very powerful tool that can do an awful lot of very neat things.

A bit of terminology about Excel before we move on. The first ever Excel-like computer application was called Visicalc (thank you, Dan Bricklin and Bob Frankston) and was known as a spreadsheet. OK—Excel is known as a spreadsheet program as well, but each individual sheet is known as a **worksheet.** And worksheets, when combined, constitute what is known as a **workbook.** Fun, huh?

A FIVE-MINUTE HISTORY OF STATISTICS

Before you read any further, it would be useful to have some historical perspective about this topic called statistics. After all, almost every undergraduate in the social, behavioral, and biological sciences and every graduate student in education, nursing, psychology, social welfare and social services, and anthropology (you get the picture) is required to take this course. Wouldn't it be nice to have some idea from whence the topic it covers came? Of course it would.

Way, way back, as soon as humans realized that counting was a good idea (as in "How many of these do you need to trade for one of those?"), collecting information also became a useful skill.

If counting counted, then one would know how many times the sun would rise in one season, how much food was needed to last the winter, and what amount of resources belonged to whom.

That was just the beginning. Once numbers became part of language, it seemed like the next step was to attach these numbers to outcomes. That started in earnest during the 17th century, when the first set of data pertaining to populations was collected. From that point on, scientists (mostly mathematicians, but then physical and biological scientists) needed to develop specific tools to answer specific questions. For example, Francis Galton (a cousin of Charles Darwin, by the way), who lived from 1822 to 1911, was very interested in the nature of human intelligence. To explore one of his primary questions regarding the similarity of intelligence among family members, he used a specific statistical tool called the correlation coefficient (first developed by mathematicians), and then he popularized its use in the behavioral and social sciences.

You'll learn all about this tool in Chapter 5. In fact, most of the basic statistical procedures that you will learn about were first developed and used in the fields of agriculture, astronomy, and even politics. Their application to human behavior came much later.

The past 100 years have seen great strides in the invention of new ways to use old ideas. The simplest test for examining the differences between the averages of two groups was first advanced during the early 20th century. Techniques that build on this idea were offered decades later and have been greatly refined. And the introduction of personal computers and such programs as Excel have opened up the use of sophisticated techniques to anyone who wants to explore these fascinating topics.

The introduction of these powerful personal computers has been both good and bad. It's good because most statistical analyses no longer require access to a huge and expensive mainframe computer. Instead, a simple personal computer costing less than $500 can do 95% of what 95% of the people need. On the other hand, less than adequately educated students (such as your fellow students who passed on taking this course!) will take any old data they have and think that by running them through some sophisticated analysis, they will have reliable, trustworthy, and meaningful outcomes—not true. What your professor would say is, "Garbage in, garbage out"; if you don't start with reliable and trustworthy data, what you'll have after your data are analyzed are unreliable and untrustworthy results.

Today, statisticians in all different areas, from criminal justice to geophysics to psychology, find themselves using basically the same techniques to answer different questions. There are, of course, important differences in how data are collected, but for the most part, the analyses (the plural of analysis) that are done following the collection of data (the plural of datum) tend to be very similar, even if called something different. The moral here? This class will provide you with the tools to understand how statistics are used in almost any discipline. Pretty neat, and all for just three or four credits.

If you want to learn more about the history of statistics and see a historical time line, a great place to start is at Saint Anselm's College Internet site, located at http://www.anselm.edu/homepage/jpitocch/biostatshist.html and http://www.stat.ucla.edu/history (at the University of California, Los Angeles).

OK. Five minutes is up and you know as much as you need to know about the history of statistics. Let's move on to what it is (and isn't).

STATISTICS: WHAT IT IS (AND ISN'T)

Statistics for People Who (Think They) Hate Statistics . . . Excel 2007 Edition is a book about basic statistics and how to apply them to a variety of different situations, including the analysis and understanding of information.

In the most general sense, **statistics** describes a set of tools and techniques that is used for describing, organizing, and interpreting information or data. Those data might be the scores on a test taken by students participating in a special math curriculum, the speed with which problems are solved, the number of patient complaints when using one type of drug rather than another, the number of errors in each inning of a World Series game, or the average price of a dinner in an upscale restaurant in Santa Fe.

In all of these examples, and the million more we could think of, data are collected, organized, summarized, and then interpreted. In this book, you'll learn about collecting, organizing, and summarizing data as part of descriptive statistics. And then you'll learn about interpreting data when you learn about the usefulness of inferential statistics.

What Are Descriptive Statistics?

Descriptive statistics are used to organize and describe the characteristics of a collection of data. The collection is sometimes called a **data set** or just **data**.

For example, the following list shows you the names of 22 college students, their major areas of study, and their ages. If you needed to describe what the most popular college major is, you could use a descriptive statistic that summarizes their choice (called the mode). In this case, the most common major is psychology. And if you wanted to know the average age, you could easily compute another descriptive statistic that identifies this variable (that one's called the mean). Both of these simple descriptive statistics are used to describe data. They do a fine job allowing us to represent the characteristics of a large collection of data such as the 22 cases in our example.

Name	Major	Age	Name	Major	Age
Richard	Education	19	Elizabeth	English	21
Sara	Psychology	18	Bill	Psychology	22
Andrea	Education	19	Hadley	Psychology	23
Steven	Psychology	21	Buffy	Education	21
Jordan	Education	20	Chip	Education	19
Pam	Education	24	Homer	Psychology	18
Michael	Psychology	21	Margaret	English	22
Liz	Psychology	19	Courtney	Psychology	24
Nicole	Chemistry	19	Leonard	Psychology	21
Mike	Nursing	20	Jeffrey	Chemistry	18
Kent	History	18	Emily	Spanish	19

So watch how simple this is. To find the most frequently selected major, just find the one that occurs most often. And to find the average age, just add up all the age values and divide by 22. You're right—the most often occurring major is psychology (9 times) and the average age is 20.3 (actually 20.27). Look, Ma! No hands—you're a statistician.

What Are Inferential Statistics?

Inferential statistics are often (but not always) the next step after you have collected and summarized data. Inferential statistics are used to make inferences from a smaller group of data (such as our group of 22 students) to a possibly larger one (such as all the undergraduate students in the College of Arts and Sciences).

This smaller group of data is often called a **sample,** which is a portion, or a subset, of a **population.** For example, all the fifth graders in Newark, New Jersey, would be a population (it's all the occurrences with certain characteristics—being in fifth grade and living in Newark), whereas a selection of 150 of them would be a sample.

Let's look at another example. Your marketing agency asks you (a newly hired researcher) to determine which of several different names is most appealing for a new brand of potato chip. Will it be Chipsters? FunChips? Crunchies? As a statistics pro (we know we're moving a bit ahead of ourselves, but keep the faith), you need to find a small group of potato chip eaters that is representative of all potato chip fans and ask these people to tell you which one of the three names they like the most. Then, if you did things right, you can easily extrapolate the findings to the huge group of potato chip eaters.

Or let's say you're interested in the best treatment for a particular type of disease. Perhaps you'll try a new drug as one alternative, a placebo (a substance that is known not to have any effect) as another alternative, and nothing as the third alternative to see what happens. Well, you find out that a larger number of patients get better when no action is taken and nature just takes its course! The drug does not have any effect. Then, with that information, you extrapolate to the larger group of patients that suffers from the disease, given the results of your experiment.

In Other Words . . .

Statistics is a tool that helps us understand the world around us. It does so by organizing information we've collected and then letting us make certain statements about how characteristics of those data are applicable to new settings. Descriptive and inferential statistics work hand in hand, and which one you use and when depends on the question you want answered.

And today, a knowledge of statistics is more important than ever because it provides us with the tools to make decisions that are based on empirical (observed) evidence and not our own biases or beliefs. Like to see if early intervention programs work? Then test whether they work and provide that evidence to the court where a ruling will be made on the viability of a new school bond issue and paying for those programs.

TOOLING AROUND WITH
THE ANALYSIS TOOLPAK

An awful lot of what you need to know about Excel can be found in Appendix A. However, there are certain Excel procedures available only if you have the Analysis ToolPak installed (and we use those tools in several chapters throughout the book).

The Analysis ToolPak is a spectacular Excel add-in, a special set of tools that may not have been installed when Excel was originally installed. How do you know if it is installed on the computer you are using? If the Analysis ToolPak doesn't appear on your Data menu (usually on the right side as you see in Figure 1.1), you need to install it. Either ask your instructor to have this done on the network level where Excel is installed, or install it on your own machine by doing the following.

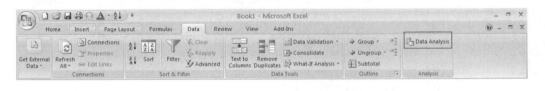

Figure 1.1   The Analysis ToolPak on the Data Tab

1. Click the Microsoft Office Button, and then click Excel Options.

2. Click Add-Ins, and then in the Manage box, select Excel Add-Ins.

3. Click Go.

4. In the Add-Ins Available box, select the Analysis ToolPak check box, and then click OK.

The Analysis ToolPak is an Excel add-in. An add-in is a program that adds custom commands and features to Microsoft Office. Microsoft Office is Excel's mother and father.

You are now done and ready to make your Excel activities even that much more productive and fun. You can learn how to use the Analysis ToolPak in Little Chapter 1b.

WHAT AM I DOING IN A STATISTICS CLASS?

There are probably many reasons why you find yourself using this book. You might be enrolled in an introductory statistics class. Or, you might be reviewing for your comprehensive exams. Or, you might even be reading this on summer vacation (horrors!) in preparation and review for a more advanced class.

In any case, you're a statistics student whether you have to take a final exam at the end of a formal course or whether you're just in it on your own accord. But there are plenty of good reasons to be studying this material—some fun, some serious, and some both.

Here's the list of some of the things that my students hear at the beginning of our introductory statistics course.

1. Statistics 101 or Statistics 1 or whatever it's called at your school looks great listed on your transcript. Kidding aside, this may be a required course for you to complete your major. But even if it is not, having these skills is definitely a big plus when it comes time to apply for a job or for further schooling. And with more advanced courses, your résumé will be even more impressive.

2. If this is not a required course, taking basic statistics sets you apart from those who do not. It shows that you are willing to undertake a course that is above average with regard to difficulty and commitment.

3. Basic statistics is an intellectual challenge of a kind that you might not be used to. There's a good deal of thinking that's required, a bit of math, and some integration of ideas and application. The bottom line is that all this activity adds up to what can be an invigorating intellectual experience because you learn about a whole new area or discipline.

4. There's no question that having some background in statistics makes you a better student in the social or behavioral sciences because you will have a better understanding not only of what you read in journals but also of what your professors and colleagues may be discussing and doing in and out of class. You will be amazed the first time you say to yourself, "Wow, I actually understand what they're talking about." And it will happen over and over again because you will have the basic tools necessary to understand exactly how scientists reach the conclusions they do.

5. If you plan to pursue a graduate degree in education, anthropology, economics, nursing, sociology, or any one of many other social, behavioral, and biological pursuits, this course will give you the foundation you need to move further.

6. Finally, you can brag that you completed a course that everyone thinks is the equivalent of building and running a nuclear reactor.

TEN WAYS TO USE THIS BOOK (AND LEARN STATISTICS AT THE SAME TIME!)

Yep. Just what the world needs—another statistics book. But this one is different. It's directed at the student, is not condescending, is informative, and is as basic as possible in its presentation. It makes no presumptions about what you should know before you start and proceeds in slow, small steps, which lets you pace yourself.

However, there has always been a general aura surrounding the study of statistics that it's a difficult subject to master. And we don't say otherwise, because parts of it are challenging. On the other hand, millions and millions of students have mastered this topic, and you can, too. Here are a few hints to close this introductory chapter before we move on to our first topic.

1. **You're not dumb.** That's true. If you were, you would not have gotten this far in school. So, treat statistics like any other new course. Attend the lectures, study the material, do the exercises in the book and from class, and you'll do fine. Rocket scientists know statistics, but you don't have to be a rocket scientist to succeed in statistics.

2. **How do you know statistics is hard?** Is statistics difficult? Yes and no. If you listen to friends who have taken the course and didn't work hard and didn't do well, they'll surely volunteer to tell you how hard it was and how much of a disaster it made of their entire semester, if not their lives. And let's not forget—we always tend to hear from complainers. So, we'd suggest that you start this course with the attitude that you'll wait and see how it is and judge the experience for yourself. Better yet, talk to several people who have had the class and get a good general idea of what they think. Just don't base it on one spoilsport's experience.

3. **Don't skip lessons—work through the chapters in sequence.** *Statistics for People Who (Think They) Hate Statistics . . . Excel 2007 Edition* is written so that each chapter provides a foundation for the next one in the book. When you are all done with the course, you will (I hope) refer back to this book and

use it as a reference. So, if you need a particular value from a table, you might consult Appendix B. Or, if you need to remember how to compute the standard deviation, you might turn to Chapter 3. But for now, read each chapter in the sequence that it appears. It's OK to skip around and see what's offered down the road. Just don't study later chapters before you master earlier ones.

4. **Form a study group.** This is one of the most basic ways to ensure some success in this course. Early in the semester, arrange to study with friends. If you don't have any who are in the same class as you, then make some new ones or offer to study with someone who looks as happy to be there as you are. Studying with others allows you to help them if you know the material better, or to benefit from others who know that material better than you. Set a specific time each week to get together for an hour and go over the exercises at the end of the chapter or ask questions of one another. Take as much time as you need. Studying with others is an invaluable way to help you understand and master the material in this course.

5. **Ask your teacher questions, and then ask a friend.** If you do not understand what you are being taught in class, ask your professor to clarify it. Have no doubt—if you don't understand the material, then you can be sure that others do not as well. More often than not, instructors welcome questions. And especially because you've read the material before class, your questions should be well informed and help everyone in class to better understand the material.

6. **Do the exercises at the end of a chapter.** The exercises are based on the material and the examples in the chapter they follow. They are there to help you apply the concepts that were taught in the chapter and build your confidence at the same time. How do the exercises do that? An explanation for how each exercise is solved accompanies the problem. If you can answer these end-of-chapter exercises, then you are well on your way to mastering the content of the chapter.

7. **Practice, practice, practice.** Yes, it's a very old joke:

   Q. How do you get to Carnegie Hall?

   A. Practice, practice, practice.

   Well, it's no different with basic statistics. You have to use what you learn and use it frequently to master the different ideas and techniques. This means doing the exercises in the back of Chapters 1 through 16 as well as taking advantage of any other opportunities you have to understand what you have learned.

8. **Look for applications to make it more real.** In your other classes, you probably have occasion to read journal articles, talk about the results of research, and generally discuss the importance of the scientific method in your own area of study. These are all opportunities to look and see how your study of

statistics can help you better understand the topics under class discussion as well as the area of beginning statistics. The more you apply these new ideas, the better and more full your understanding will be.

9. **Browse.** Read over the assigned chapter first, then go back and read it with more intention. Take a nice leisurely tour of *Statistics for People Who (Think They) Hate Statistics* to see what's contained in the various chapters. Don't rush yourself. It's always good to know what topics lie ahead as well as to familiarize yourself with the content that will be covered in your current statistics class.

10. **Have fun.** This might seem like a strange thing to say, but it all boils down to you mastering this topic rather than letting the course and its demands master you. Set up a study schedule and follow it, ask questions in class, and consider this intellectual exercise to be one of growth. Mastering new material is always exciting and satisfying—it's part of the human spirit. You can experience the same satisfaction here—just keep your eye on the ball and make the necessary commitment to stay current with the assignments and work hard.

And a short note for Mac users. Over the years, the Excel people at Microsoft have become increasingly kind to users of the Macintosh version. At this point, the latest versions of Excel for a Windows operating system and a Macintosh operating system are almost identical. The biggest difference (and it really isn't very big) are the keystrokes that one uses to accomplish particular tasks. So, for example, instead of using the Ctrl+C key combination to copy highlighted text windows, the Mac uses the **Apple** or the **Command key** (the cool little key on the lower left of the keyboard with the famous Apple logo and four little squiggles) in combination with the C key to accomplish the same. This Apple key is also referred to (believe it or not) as the splat, the cloverleaf, the butterfly, the beanie, and the flower key. Using Excel in one operating system or the other (or both) requires a very similar set of tasks, and you should have no problem making the adjustment. All that said, if you really want to impress your friends, Mac users can go to System Preferences, and reconfigure the keyboard to ensure that Windows and Mac keystrokes are exactly the same!

ABOUT THOSE ICONS

An icon is a symbol. Throughout *Statistics for People . . .* , you'll see a variety of different icons. Here's what each one is and what each represents:

This icon represents information that goes beyond the regular text. We might find it necessary to elaborate on a particular point, and we can do that more easily outside of the flow of the usual material.

TECH TALK Here, we select some more technical ideas and tips to discuss and to inform you about what's beyond the scope of this course. You might find these interesting and useful.

Throughout *Statistics for People . . .* , you'll find a small stepladder icon like the one you see here. This indicates that there is a set of steps coming up that will direct you through a particular process. These steps have been tested and approved by whatever federal agency approves these things.

That finger with the bow is a cute icon, but its primary purpose is to help reinforce important points about the topic that you just read about. Try to emphasize these points in your studying, because they are usually central to the topic.

Many of the chapters in *Statistics for People . . .* provide detailed information about one or more particular statistical procedures and the computation that accompanies them. The computer icon is used to identify the "Using the Computer to . . ." section of the chapter.

The More Excel icon identifies additional information on the Excel feature that has just been mentioned or worked with.

Appendix A, Excel-erate Your Learning: All You Need to Know About Excel, contains a collection of 50 basic and important tasks that anyone who uses Excel should know.

Appendix B contains important tables you will learn about and need throughout the book.

And, in working through the exercises in this book, you will use the data sets in Appendix C, starting on page 366. You'll find reference to data sets (such as "Chapter 2 Data Set 1"), and each of these sets is shown in Appendix C. You can either enter the data manually or download them from the book's Internet site at www.sagepub .com/salkind2datasets. Just click on Data Sets and save it. Or, you can download the data sets from the author's own Web site at the University of Kansas (http://soe.ku.edu/faculty/Salkind/

statsexcel_fpwhs). Click on *Statistics for People . . . Excel 2007 Edition* to get to the book's home page, and then download the Excel files.

KEY TO DIFFICULTY ICONS

To help you along a bit, we placed a difficulty index at the beginning of each chapter. This adds some fun to the start of each chapter, but it's also very useful as a tip to let you know what's coming and how difficult chapters are in relation to one another.

☺ (very hard)

☺☺ (hard)

☺☺☺ (not too hard, but not easy either)

☺☺☺☺ (easy)

☺☺☺☺☺ (very easy)

KEY TO "HOW MUCH EXCEL" ICONS

To help you along a bit more, we placed a "How Much Excel" index at the beginning of each chapter. This adds even more fun (groan) to the start of each chapter, but also lets you know how much Excel material is contained in the chapter.

How much Excel?

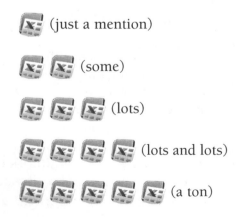

(just a mention)

(some)

(lots)

(lots and lots)

(a ton)

GLOSSARY

Bolded terms in the text are included in the glossary at the back of the book.

Summary

That couldn't have been that bad, right? We want to encourage you to continue reading and not worry about what's difficult or time consuming or too complex for you to understand and apply. Just take one chapter at a time, as you did this one.

Time to Practice

Because there's no substitute for the real thing, Chapters 1 through 16 end with a set of exercises that will help you review the material that was covered in the chapter. And so you don't have to go hunting, the answers to these exercises can be found at the very end of those chapters as well.

For example, here is the first set of exercises (but don't look for any answers for these because these are kind of "on your own" answers—each answer is highly tied to your own experiences and interest).

1. Interview someone who uses statistics in his or her everyday work. It might be your adviser, an instructor, a researcher who lives on your block, a market analyst for a company, or even a city planner. Ask the person what his or her first statistics course was like. Find out what he or she liked and didn't like. See if he or she has any suggestions to help you succeed. And most important, ask the person about the ways he or she uses these new-to-you tools at work.

2. Search through your local newspaper and find the results of a survey or interview about any topic. Summarize what the results are, and do the best job you can describing how the researchers who were involved, or the authors of the survey, came to the conclusions they did. It may or may not be apparent. Once you have some idea of what they did, try to speculate as to what other ways the same information might be collected, organized, and summarized.

3. Go to the library and copy a journal article in your own discipline. Then, go through the article with one of those fancy highlighters and highlight the section (usually the "Results" section) where statistical procedures were used to organize and analyze the data. You don't know much about the specifics of this yet, but

how many of these different procedures (such as *t* test, mean, and calculation of the standard deviation) can you identify? Can you take the next step and tell your instructor how the results relate to the research question or the primary topic of the research study?

4. Find five Web sites on the Internet that contain data on any topic, and write a brief description of what type of information is offered and how it is organized. For example, if you go to the mother of all data sites, the U.S. Census (at http://www.census.gov), you'll find links to Access Tools, which takes you to a page just loaded with links to real live data. Try to find data and information that fit in your own discipline.

5. And the big extra credit assignment is to find someone who actually uses Excel for his or her daily data analysis needs. Ask why he or she uses Excel rather than a more singularly focused program such as SPSS. You may very well find these good folks in everything from political science to nursing, so search widely!

1A

All You Need to Know About Formulas and Functions

Difficulty Scale ☺☺ (a little tough, but invaluable to stick with)

How much Excel? 🅧 🅧 🅧 🅧 (lots and lots)

What you'll learn about in this chapter

- The difference between formulas and functions
- How to create and use a formula
- The important Excel functions
- How to select and use a function

There may be nothing more valuable in your Excel magic tool box than formulas and functions. They both allow you to bypass (very) tedious calculations and get right to the heart of the matter. Both formulas and functions are short cuts—and both work in different ways and do different things. Let's start with formulas.

WHAT'S A FORMULA?

You probably already know the answer to that question. A **formula** is a set of mathematical operators that performs a particular mathematical task. For example, here's a simple formula:

$$2 + 2 =$$

The operator "+" tells you to add certain values (a 2 and another 2) together to produce the outcome 4. This is a simple one.

Here's one that's a bit more advanced and one with which you will become more familiar in Chapter 14 of *Statistics for People . . .* :

$$Y' = bX + a. \tag{1a.1}$$

This is the formula that is used to predict the value of *Y* from our knowledge of the values of *b*, *X*, and *a*. We'll worry about what all those mean later, but it's a formula that contains a bunch of symbols and mathematical operators and helps us compute numbers we need to make decisions.

Excel is a formula engine just ready for you to use these tools to make your learning of statistics easier.

Creating a Formula

A formula is created through these steps.

1. Click on the **cell** in which you want the results of the formula to appear.

2. Enter an equal sign, which looks like this: =.

   All formulas begin with an equal sign, no matter what else they contain.

3. Enter the formula. No spaces in formulas please—Excel does not like them.

4. Press the Enter key, and voilà, the results of the formula will appear in the selected cell.

For example, let's enter the formula that was shown earlier—2+2—and see how these steps work.

1. As you can see in Figure 1a.1, we selected Cell A1.

A1		▼	*fx*
A	**B**	**C**	**D**
1			
2			
3			

Figure 1a.1   Selecting a Cell Into Which a Formula Will Be Entered

2. The equal sign is entered, as shown in Figure 1a.2. And, as you can see, the formula bar at the top of the column becomes active. Everything we enter in Cell A1 will appear in the formula bar.

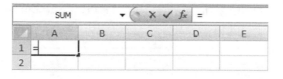

Figure 1a.2 Entering the Equal Sign to Indicate the Beginning of a Formula

3. Enter the rest of the formula, which in this case is 2+2, as you see in Figure 1a.3.

Figure 1a.3 Entering the Formula in Cell A1

4. Press Enter, and the value of the formula is returned to the cell, as you see in Figure 1a.4.

Figure 1a.4 The Value of a Formula Being Returned to the Cell

A few notes:

- A formula always begins with an equal sign, which tells Excel that what follows is the formula.
- The formula itself always appears in the formula bar.
- The results of the formula (and not the formula itself) are returned to the selected cell.

This is the simplest example of how to use a formula, and formulas can become as complex as you need them to be.

More Excel

Want to see the formula behind the screens in a worksheet? Just use the Ctrl+` key combination to toggle between formulas and the results of those formulas. The ` key is to the left of the number 1 key on the left-hand side of the keyboard.

Operator, Operator—Get Me a Formula!

You have just seen that even the most simple formulas consist of operators, and in this simple case, the only one is a plus ("+") sign, which directs Excel to add the two values you see in Figure 1a.3 together and return the sum to Cell A1.

Adding is just one type of operation you can perform. The most important operations and the symbols you use to accomplish them are shown in the following table.

Operator	Symbol	Example	What It Does
Addition	+ (plus)	=2+5	Adds 2 and 5
Subtraction	– (minus)	=5–3	Subtracts 3 from 5
Division	/ (slash)	=10/5	Divides 10 by 5
Multiplication	* (asterisk)	=2*5	Multiplies 2 times 5
Power of	^ (caret)	=4^2	Takes 4 to the power of 2, or squares 4

Beware the Parentheses

When you create a formula that goes beyond a very simple one, it is critical for you to consider the order in which operations are carried out and the use of parentheses.

Let's say that we want to find the average score on a weekly test given each Friday for a month and the scores range from 0 to 100.

Here are Willy's scores:

Week 1   78

Week 2   56

Week 3   85

Week 4   92 (Willy finally got it!)

We need to create a formula that will add each of the scores together and divide the sum by 4. We'll name each score w_1, w_2, w_3, etc. Here's one way to do it:

$$w_1 + w_2 + w_3 + w_4/4.$$

Oops—what this does is it adds w_1, w_2, and w_3 together, and adds that sum to the value of w_4 divided by 4. This is a no-no and not what we want.

Rather, take a look at this formula:

$$(w_1 + w_2 + w_3 + w_4)/4.$$

This is what we want. Here, the four values were summed and then divided by four. This one works. The lesson? Watch your parentheses when you have anything more than the most simple equation.

WHAT'S A FUNCTION?

You know that a formula is a simple set of symbols (such as numbers and operators) that performs some calculation and results in an outcome in the cell where the formula lives.

A **function** is nothing other than a predefined formula. The good people who created Excel developed a whole bunch of these that can do many different things, but for our purposes throughout *Statistics for People . . .* , we deal only with those that are relevant to the chapters of this book. These functions fall under the general Formulas tab on the ribbon as you see in Figure 1a.5.

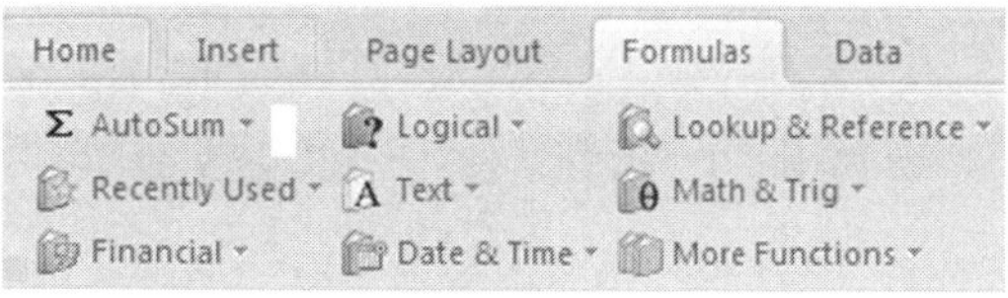

Home	Insert	Page Layout	Formulas	Data
Σ AutoSum ▾		Logical ▾	Lookup & Reference ▾	
Recently Used ▾	A Text ▾	Math & Trig ▾		
Financial ▾	Date & Time ▾	More Functions ▾		

Figure 1a.5   The Formulas Tab on the Excel Ribbon

For example, there are groups of mathematical functions, database functions, and logical functions. But we're going to focus only on the functions that fall in the category of statistical functions (which may be visible on your screen only if you click on the More

Functions drop-down box, as you see in Figure 1a.5, and then click Statistical).

And, out of all of these statistical functions, we deal only with the ones that are relevant to the material covered in this book, such as **AVERAGE** (guess what that does) and **TTEST** (guess, but you probably don't yet know). Some are too advanced for us to bother with, such as FOURIER and FISHER. We'll leave those for the next course or for you to explore on your own.

Using a Function

Unlike a formula, a function is not created by you. You just tell it which values (in which cells) you want to work with. And, every formula contains two different elements—the name of the function and the argument of the function.

Here's a very simple function that averages a given set of numbers. In this example, this function averages the numbers in Cells A1 through A3:

=AVERAGE(A1:A3).

The name of the function is AVERAGE, and the argument is A1:A3—the cells on which you want the function to perform its magic. And as you can see, functions (like formulas) always, always, always begin with an equal sign.

Here's another function that produces the sum of a set of cells:

=SUM(A1:A3).

Simple, right? And, you may be thinking, "Well, why not just use a formula in this case?" and you could. But what if you needed the sum of a set of 3,267 values like this:

=SUM(A1:A3267)?

Or a very fancy-schmancy calculation that includes functions that are very complex? Those are both different stories, and you really didn't want to type in =(A1+A2+A3+A4 . . .) until you get to A3267, right? We thought not.

So, let's get to the way that we use a function, and as an example, we'll use the SUM function.

To use this (or any other function), you follow three steps.

1. Enter the function in the cell where you want the results to appear.

2. Enter the range of cells on which you want the function to operate.

3. Press the Enter key, and voilá, there you have it.

However, there are several ways to accomplish these three steps, and let's deal with those now.

Inserting a Function (When You Know the Function's Name and How It Works)

OK, here's the old-fashioned way.

1. Enter the function in the cell where you want the results to appear.

   For example, in Figure 1a.6, you can see a data set of 10 values. We are going to sum those values using the SUM function. And, to make things a bit clearer, we entered a text label in the cell to the left of where we want the sum to appear.

	A	B
1		Value
2		3
3		4
4		2
5		3
6		4
7		5
8		4
9		3
10		2
11		3
12	Sum of Values	

Figure 1a.6   Creating a Data Set and the Location of the SUM Function

2. Type =SUM(B2:B11) in Cell B12.

3. Press the Enter key, and voilá, as you see in Figure 1a.7, the sum shows up in Cell B12, and in the formula bar, you can see the structure of the function.

| B12 | ▼ | ● | f_x | =SUM(B2:B11) |
A	B	C	D	E
1	Value			
2	3			
3	4			
4	2			
5	3			
6	4			
7	5			
8	4			
9	3			
10	2			
11	3			
12 Sum of Values	33			

Figure 1a.7   The Completed SUM Function

Notice that the results of the function (33) are returned to the same cell (B12) where the function was entered. Pretty cool.

And, not very difficult and very convenient. Remember that you can do this with any function. But how do you know what the structure of the function is? That's where the next step comes in.

More Excel

OK—so how do you know what function to use? Well, certainly one way is through exploring different functions and finding out what they do (which you will do throughout *Statistics for People.* . . . Another is by using Excel Help (press F1 at any time and enter the terms on which you want help). And, another way is to look at Table 1a.1 at the end of this little chapter, which gives you a heads-up on which functions we'll be mentioning (some in great detail and others just in passing) throughout the book and what they do.

Inserting a Function Using the Insert Command

Let's use the same example, the SUM function, and assume you haven't used it before but know this is the one you want to use.

We're using the same data as shown in Figure 1a.6.

1. Select Cell B12.

2. Click the Formula tab and then the Insert Function option on the left side of the ribbon. When you do this, you will see the Insert Function dialog box as in Figure 1a.8.

Figure 1a.8 The Insert Function Dialog Box

3. Now you can do one of two things.

   a. Type a brief description for what you want to do, such as "SUM"

   or

   b. Find the function you want in the list of functions and double click on it. This is the one we did, and when we did, the Function arguments dialog box appeared, as shown in Figure 1a.9.

Function Arguments

SUM

Number1   `32:B11`   = {3;4;2;3;4;5;4;3;2;3}
Number2   = number

                                      = 33

Adds all the numbers in a range of cells.

Number1: number1,number2,... are 1 to 255 numbers to sum. Logical values and text
are ignored in cells, included if typed as arguments.

Formula result = 33

Help on this function   OK   Cancel

Figure 1a.9   The Function Arguments Dialog Box

Don't get too excited. A function's argument is not really an argument like a disagreement. An argument in mathematical terms is a
set of premises, and that's exactly what you need to provide within
the parentheses of any function—a set of premises that the function is to carry out.

Let's take a look at the different elements in this dialog box.

- There's the name of the function, SUM.
- Then there are text boxes where you enter the range of cells (the
  argument) on which you want the function to perform its duty.
  Note that Excel is pretty smart and automatically entered the
  range of cells it thinks you want to sum. Notice that the actual
  numbers you want to sum are listed to the right of the text box.
- In the middle of the dialog box (on the left) and to the right of that
  is the value the function will return to the cell in which it is
  located (which in this case is 33).
- The syntax (or directions) of how to put the function together.
- The formula result, and
- Finally, a place to get help if you need it.

4. Click OK, and you will see the same result as you see in Figure 1a.6.
   We just entered the function using the Insert command, but we get
   the same result.

More Excel

Most functions can do a lot more than first appears. For example, with the most simple of functions such as SUM, you can enter the following variations and get the following results. Excel functions are so useful because they are so flexible.

If you enter the following formula . . .	Excel does this . . .
=SUM(3,4)	Adds the values to get 7
=SUM(A2:A4)	Adds the values located in Cells A2 through A4
=SUM(A2:A4,6)	Adds the values located in Cells A2 through A4 and also adds the value of 6 to that sum
=SUM(A6:A8,4)	Adds the values located in Cells A6 to A8 and adds the value of 4 to that sum

More Excel

Now you know two ways to insert a function in a worksheet—by typing its name or selecting it through the Insert Function dialog box. And once the specific Function Arguments dialog box (like the one you see in Figure 1a.9) is open, you can just enter the cell addresses in the appropriate text box. However, you can also just click in the cell address box and then drag the mouse over the cell addresses you want to include in that box. Good. But there's another nifty way to help you. You can click on the Collapse button (which looks like this: 🖼 which will shrink the entire dialog box and allow you to select the cells you want using the mouse directly on the worksheet. Then click the Expand button 🖼 and the dialog box returns to its normal size with the cell addresses included.

Using Functions in Formulas

Time to get a bit fancy.

Now, formulas and functions are basically the same animal—they carry out instructions. There's just no reason why you can't include a function in a formula.

For example, let's say that you have three evaluation scores (Eval 1, Eval 2, and Eval 3) as you see in Figure 1a.10. You also have a Fudge Factor (in column E), and that's a value you can use to increase or decrease an employee's score at your discretion. For example, for employee KH, you want to increase that score by 3%, so you multiply the average evaluation score (from Eval 1, Eval 2, and Eval 3) by 1.03. Here's the formula using the AVERAGE function (which you will learn how to use in Chapter 2).

	F2			▼	f_x =AVERAGE(B2:D2)*E2	
	A	B	C	D	E	F
1	Name	Eval 1	Eval 2	Eval 3	Fudge Factor	Final Score
2	HY	67	76	34	1.10	64.90
3	NM	65	78	32	1.02	59.50
4	GG	45	98	34	1.23	72.57
5	DF	65	98	54	1.06	76.67
6	KH	76	78	58	1.03	72.79
7	RR	32	76	75	1.04	63.44
8	YR	45	77	54	1.05	61.60
9	HH	43	76	54	1.00	57.67
10	JU	34	54	33	1.01	40.74
11	WE	32	47	78	1.04	54.43

Figure 1a.10   Using a Function in a Formula

As you can see in the formula bar shown in Figure 1a.10, the formula looks like this:

$$=AVERAGE(B2:D2)*E2$$

and reads like this: The contents of Cells B2 through D2 are averaged and then that value is multiplied by the contents of Cell E2. We copied the formula from Cell F2 through Cell F11, and the results are shown in Column F.

We're Taking Names: Naming Ranges

It's certainly easy enough to enter cell addresses such as A1:A3—not much work involved there.

But what if you're dealing with a really large worksheet where there are hundreds of columns and rows and thousands of cells? Wouldn't it be nice if you could just enter a name that represents a certain range of cells rather than having to remember all those cell addresses? Desire it no more. Excel allows you to name a **range,** or a collection of cells.

For example, in Figure 1a.10, instead of using the cell addresses C2:C11, why not just give the range of cells a name, such as eval2 or EVAL_2 (no spaces, please). Then, the average for that set of scores using the AVERAGE function would look like this

=AVERAGE(EVAL_2)

rather than

=AVERAGE(C5:C14).

And it gets better—you can just paste that name into any formula or function with a few clicks. You don't even have to type anything! Here's how to assign a name to a range of cells:

1. Highlight the range of cells you want to name.

2. Click the Name box at the left end of the formula bar.

3. Type the name that you want to use to refer to your selection as shown in Figure 1a.11.

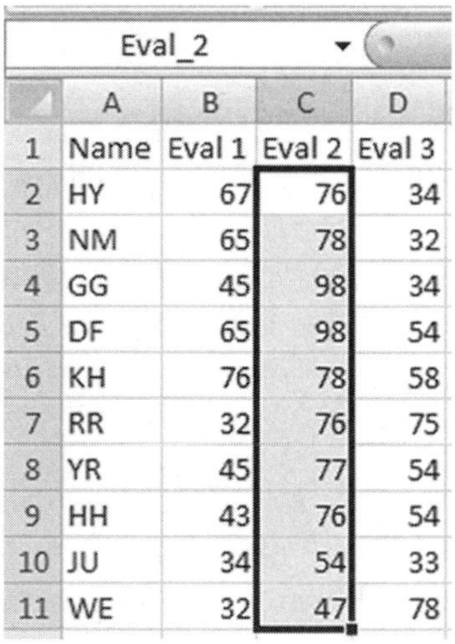

	A	B	C	D
	Name	Eval 1	Eval 2	Eval 3
1				
2	HY	67	76	34
3	NM	65	78	32
4	GG	45	98	34
5	DF	65	98	54
6	KH	76	78	58
7	RR	32	76	75
8	YR	45	77	54
9	HH	43	76	54
10	JU	34	54	33
11	WE	32	47	78

Figure 1a.11   Naming a Selection of Cells as a Range

4. Press Enter.

Using Ranges

Once a range is defined, the name you assign can be used instead of a cell range. If you remember that they are named, you can just enter that, but take a look at Figure 1a.12. Here, all five columns as named are shown in the Name Manager dialog box.

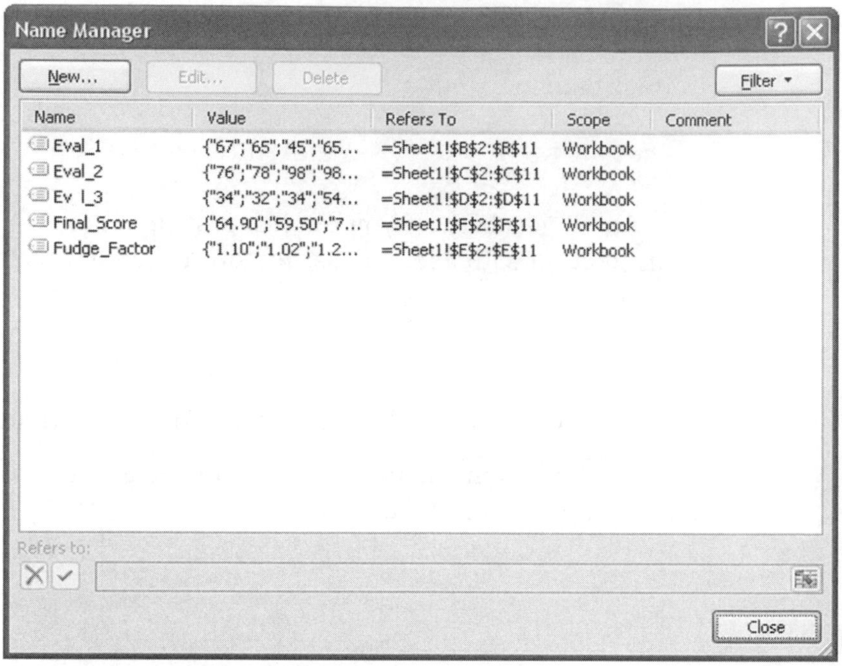

Figure 1a.12   Naming Cell Ranges

Let's use the ranges that were defined and compute the average of all the Fudge Factor scores.

1. Click on Cell F12 where the average will be placed.

2. Type =AVERAGE(.

3. Click Formulas → Use in Formula, and you will see the Use in Formula drop-down menu as shown in Figure 1a.13.

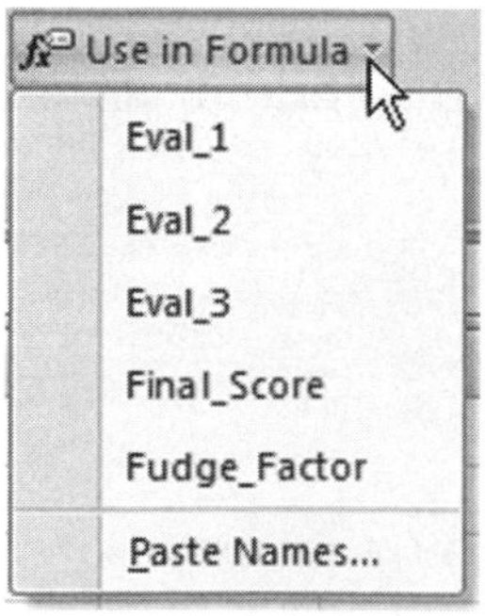

Figure 1a.13   The Menu for Selecting a Range of Cells

4. Click on Fudge_Factor.

5. Type).

6. Press Enter, and take a look at Figure 1a.14, where you can see how the name was used in the function rather than the cell address of F2 and F11.

	A	B	C	D	E	F
	Name	Eval 1	Eval 2	Eval 3	Fudge Factor	Final Score
1	Name	Eval 1	Eval 2	Eval 3	Fudge Factor	Final Score
2	HY	67	76	34	1.10	64.90
3	NM	65	78	32	1.02	59.50
4	GG	45	98	34	1.23	72.57
5	DF	65	98	54	1.06	76.67
6	KH	76	78	58	1.03	72.79
7	RR	32	76	75	1.04	63.44
8	YR	45	77	54	1.05	61.60
9	HH	43	76	54	1.00	57.67
10	JU	34	54	33	1.01	40.74
11	WE	32	47	78	1.04	54.43
12						1.06

F12   fx =AVERAGE(Fudge_Factor)

Figure 1a.14   Inserting a Cell Range Into a Function

Summary

This may be a "little" chapter, but it contains some of the most useful tools that Excel has to offer. In fact, the use of formulas and functions are really limited only by your imagination. As you use Excel, you will find more and more ways to really make these powerful tools work exactly for you.

Time to Practice

1. Create formulas for the following in an Excel worksheet:
   a. Add the values of 3 and 5 to one another.
   b. Subtract the value of 5 from 10 and multiply the outcome by 7.
   c. Average the values 5, 6, 7, and 8.

2. What would the function look like for summing the values in cells A1 through A5?

3. What would the function look like for averaging the values in cells A1 through A5?

Answers to Practice Questions

1. Take a look at Figure 1a.15 for the answers.

2. =SUM(A1:A5)

3. =AVERAGE(A1:A5)

	A	B
1	Formula	Results
2	=3+5	8
3	=(10-5)*7	35
4	=(5+6+7+8)/4	6.5
5	=3^2+4^2+5^2	50

Figure 1a.15

TABLE 1a.1 The Functions You'll Love to Love

The Function Name	What It Does	The Chapter in Which You'll Read About It
AVERAGE	Returns the average of its arguments	2
GEOMEAN	Returns the geometric mean	2
MEDIAN	Returns the median of the given numbers	2
MODE	Returns the most common value in a data set	2
QUARTILE	Returns the quartile of a data set	2
STDEV	Estimates standard deviation based on a sample	3
STDEVP	Calculates standard deviation based on the entire population	3
VAR	Estimates variance based on a sample	3

The Function Name	What It Does	The Chapter in Which You'll Read About It
VARP	Calculates variance based on the entire population	3
KURT	Returns the kurtosis of a data set	4
SKEW	Returns the skewness of a distribution	4
CORREL	Returns the correlation coefficient between two data sets	5 and 14
PEARSON	Returns the Pearson product moment correlation coefficient	5
NORMSDIST	Returns the standard normal cumulative distribution	8
STANDARDIZE	Returns a normalized value	8
TDIST	Returns the Student's t distribution	12
TTEST	Returns the probability associated with a Student's t test	12
FDIST	Returns the F probability distribution	12
FTEST	Returns the result of an F test	12
FORECAST	Returns a value along a linear trend	15
FREQUENCY	Returns a frequency distribution as a vertical array	15
LINEST	Returns the parameters of a linear trend	15
SLOPE	Returns the slope of the linear regression line	15
STEYX	Returns the standard error of the predicted y value for each x in the regression	15
TREND	Returns values along a linear trend	15
CHIDIST	Returns the one-tailed probability of the chi-square distribution	16
CHITEST	Returns the test for independence	16

1B

All You Need to Know About Using the Amazing Analysis ToolPak

Difficulty Scale ☺☺ (a little tough, but invaluable to stick with)

How much Excel? ▦ ▦ ▦ ▦ ▦ (a ton)

What you'll learn about in this chapter

- What the Analysis ToolPak is and what it does

(Almost) everything you need to know about Excel, you can learn in Appendix A. But there are certain Excel procedures available only if you have the Analysis ToolPak (which used to be called the Data Analysis ToolPak) installed (and we use those tools in several chapters throughout the book). Excel refers to this set of tools as the Analysis ToolPak, but you will see it on your screen as Data Analysis Tools or Data Analysis—no worries. The Analysis ToolPak is an Excel add-in—a special set of tools that may not have been installed when Excel was originally installed.

How do you know if it is installed on the computer you are using? If the Data Analysis item doesn't appear on the Data menu (usually at the right-hand side of the Data tab), you need to install the Analysis ToolPak. Either ask your instructor to have this done on the network level where Excel is installed, or install it on your own machine as previously discussed.

The Analysis ToolPak is easy to use. You just follow the instructions and identify the analysis you want to perform and the data on which you want it performed, and you're done. Throughout *Statistics for People . . .* , we will show you in detailed steps how this is done.

A LOOK AT THE ANALYSIS TOOLPAK

For now, let's just look at some sample output where we took a random sample of five numbers from a group of 25 numbers. In Figure 1b.1, you see the data in column A; the sample of five numbers in column C; and the dialog box that we used from the Analysis ToolPak to tell Excel what to sample, how many to sample, and where to put the results. Much more about this later in the book, but we thought you'd like to see how this cool set of tools works.

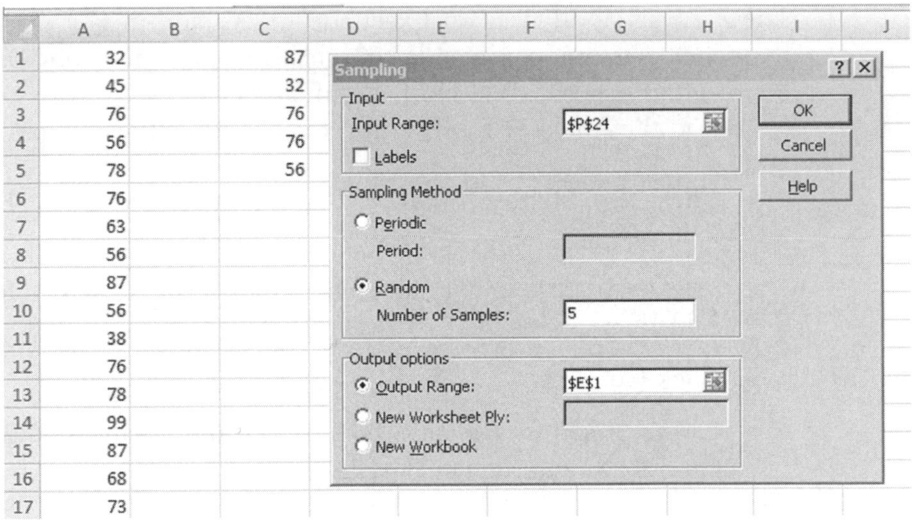

Figure 1b.1   Using the Sampling Tool From the Analysis ToolPak

For our purposes, we will be working with the following Analysis ToolPak tools (which is about 75% of them; the others are more advanced than what we need). And in the appropriate chapter (such as Chapter 2), you will learn about that particular tool (such as Descriptive Statistics):

ANOVA

Correlation

Descriptive Statistics

Histogram

Moving Average

Random Number Generation

Rank and Percentile Regression

Sampling

t-Test

z-Test

DON'T HAVE IT?

The Analysis ToolPak is an Excel add-in. An add-in is a program that adds custom commands and features to Microsoft Office. Microsoft Office is Excel's mother and father.

To load the Analysis ToolPak into Excel, follow these steps.

1. Click the Microsoft Office Button, and then click Excel Options.

2. Click Add-Ins, and then in the Manage box, select Excel Add-Ins.

3. Click Go.

4. In the Add-Ins available box, select the Analysis ToolPak check box, and then click OK.

And you are done and ready to make your Excel activities even that much more productive and fun.

Σigma Freud and Descriptive Statistics

Snapshots

And you thought *your* statistics professor was tough.

One of the things that Sigmund Freud, the founder of psycho-analysis, did quite well was to observe and describe the nature of his patients' conditions. He was an astute observer and used his skills to develop what was the first systematic and comprehensive theory of personality. Regardless of what you may think about the validity of his ideas, he was a good scientist. Back in the early 20th century, courses in statistics (like the one you are taking) were not offered as part of undergraduate or graduate curricula. The field was relatively new, and the nature of scientific explorations did not demand the precision that this set of tools brings to the scientific arena.

But things have changed. Now, in almost any endeavor, numbers count (as Francis Galton, the inventor of correlation and a first cousin to Charles Darwin, said as well). This section of *Statistics for People Who (Think They) Hate Statistics . . . Excel 2007 Edition* is devoted to how we can use Excel's most basic statistical functions to describe an outcome and better understand it.

Chapter 2 discusses measures of central tendency and how computing one of several different types of averages gives you the one best **data point** that represents a set of scores. Chapter 3 completes the coverage of tools we need to fully describe a set of data points in its discussion of variability, including the standard deviation and variance. When you get to Chapter 4, you will be ready to learn how distributions, or sets of scores, differ from one another and what this difference means. Chapter 5 deals with the nature of relationships between variables—namely, correlations—and the last of the chapters in this part, Chapter 6, focuses on reliability and validity, two topics that are critical in evaluating the assessment tools used in any research setting.

When you finish Part II, you'll be in excellent shape to start understanding the role that probability and inference play in the social and behavioral sciences.

2 Computing and Understanding Averages

Means to an End

Difficulty Scale ☺☺☺☺ (moderately easy)

How much Excel? (a ton)

What you'll learn about in this chapter

- Understanding measures of central tendency
- Computing the mean for a set of scores using the AVERAGE function
- Computing the mode for a set of scores using the MODE function
- Computing the median for a set of scores using the MEDIAN function
- Using the Analysis ToolPak to compute descriptive statistics
- Selecting a measure of central tendency

You've been very patient, and now it's finally time to get started working with some real, live data. That's exactly what you'll do in this chapter. Once data are collected, a usual first step is to organize the information using simple indexes to describe the data. The easiest way to do this is through computing an average, of which there are several different types.

An **average** is the one value that best represents an entire group of scores. It doesn't matter whether the group of scores is the number correct on a spelling test for 30 fifth graders or the batting percentage of each of the New York Yankees or the number of people who registered as Democrats or Republicans in the most recent election. In all of these examples, groups of data can be summarized using an average. Averages, also called **measures of central tendency,** come in three flavors: the mean, the median, and the mode. Each provides you with a different type of information about a distribution of scores and is simple to compute and interpret.

COMPUTING THE MEAN

The **mean** is the most common type of average that is computed. It is simply the sum of all the values in a group divided by the number of values in that group. So if you had the spelling scores for 30 fifth graders, you would simply add up all the scores and get a total and then divide by the number of students, which is 30. The formula for computing the mean is shown in Formula 2.1:

$$\bar{X} = \frac{\Sigma X}{n} \tag{2.1}$$

where

- The letter $\bar{X}$ with a line above it (also called "X bar") is the mean value of the group of scores or the mean.
- The Σ, or the Greek letter sigma, is the summation sign, which tells you to add together whatever follows it.
- The X is each individual score in the group of scores.
- Finally, the n is the size of the sample from which you are computing the mean.

To compute the mean, follow these steps:

1. List the entire set of values in one or more columns. These are all the Xs.

2. Compute the sum or total of all the values.

3. Divide the total or sum by the number of values.

For example, if you needed to compute the average number of shoppers at three different locations, you would compute a mean for that value.

Location	Number of Annual Customers
Lanham Park store	2,150
Williamsburg store	1,534
Downtown store	3,564

The mean or average number of shoppers in each store is 2,416. Formula 2.2 shows how it was computed using the formula you saw in Formula 2.1:

$$\bar{X} = \frac{\Sigma X}{n} = \frac{2,150 + 1,534 + 3,564}{3} = \frac{7,248}{3} = 2,416. \qquad (2.2)$$

See, we told you it was easy. No big deal.

And Now . . . *Using Excel's AVERAGE Function*

To compute the mean of a set of numbers using Excel, follow these steps.

> For some reason, the people who name functions would rather call the one that computes the mean AVERAGE, rather than MEAN. Yikes—these same folks used the name MEDIAN to name the function that computes the median, and they assigned the name MODE to the function that computes the mode, so why not make everyone's life easier and assign the name MEAN to the function that computes the average? If you find out, let us know.

1. Enter the individual scores into one column in a worksheet, such as you see in Figure 2.1.

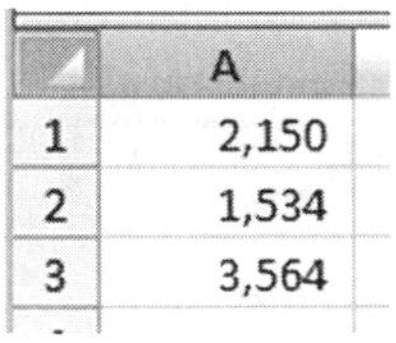

	A
1	2,150
2	1,534
3	3,564

Figure 2.1   Data That Will Be Used to Compute an Average Score

2. Select the cell (an intersection of a row and a column in a workbook) into which you want to enter the AVERAGE function. In this example, we are going to compute the mean in Cell A5.

3. Now, create a formula in any cell that would average the three values. The formula would look like this:

=(A1+A2+A3)/3,

or

click on Cell A5 and type the AVERAGE function (which we did) as follows:

=AVERAGE(A1:A3) and press the Enter key,

or

use the Formulas → Insert Function menu option and the "Inserting a Function" technique we talked about on pages 25 through 29 in Chapter 1 to enter the AVERAGE function in Cell A5.

> Whether you type in a function or enter it using the Formula → Insert Function option, it looks the same and no one will ever, ever, ever know how you did it. Once it's there, whether typed or inserted, it does exactly the same thing.

As you can see in Figure 2.2, the mean was computed and the value returned to Cell A5. Notice that in the **formula bar** (where you can see the contents of a cell) in Figure 2.2, you can see the AVERAGE function fully expressed and the value computed as 2,416, just as we did manually earlier in the chapter.

A5	▾	f_x	=AVERAGE(A1:A3)

	A	B	C	D	E
1	2,150				
2	1,534				
3	3,564				
4					
5	2,416				
6					

Figure 2.2 Using the AVERAGE Function to Compute the Mean of a Set of Numbers

More Excel

You may also want to explore the geometric mean (the function is **GEOMEAN**). The geometric mean uses multiplication rather than addition to summarize data values. It's used when one expects that changes occur in a relative fashion in the data.

THINGS TO REMEMBER

The mean is sometimes represented by the letter *M* and is also called the typical, average, or most central score. If you are reading another statistics book or a research report, and you see something like *M* = 45.87, it probably means that the mean is equal to 45.87.

- In the formula, a small *n* represents the sample size for which the mean is being computed. A large *N* (like this) would represent the population size. In some books and in some journal articles, no distinction is made between the two.
- The sample mean is the measure of central tendency that most accurately reflects the population mean.
- The mean is like the fulcrum on a seesaw. It's the centermost point where all the values on one side of the mean are equal in weight to all the values on the other side of the mean.
- Finally, for better or worse, the mean is very sensitive to extreme scores. An extreme score can pull the mean in one direction or another and make it less representative of the set of scores and less useful as a measure of central tendency. This, of course, all depends on the values for which the mean is being computed. More about this later.

TECH TALK The mean that we just computed is also referred to as the **arithmetic mean,** and there are other types of means that you may read about, such as the harmonic mean. Those are used in special circumstances but need not concern you here. And if you want to be technical about it, the arithmetic mean (which is the one that we have discussed up to now) is also defined as the point at which the sum of the deviations is equal to zero (whew!). So, if you have scores like 3, 4, and 5 (where the mean is 4), the sum of the deviations about the mean (–1, 0, and +1) is 0.

More Excel

In addition to the arithmetic mean (one of many other kinds of averages), here's also the moving average, brought to you by the Analysis ToolPak. Using the Moving Average tool, you can compute the average of a set of scores in chunks. For example, say you have the numbers 1,

4, 5, and 10. The average of these is 5.0. But a moving average taking two measures at a time (the average of 1 + 4, the average of 4 + 5, and the average of 5 + 10) equals 4.88. The moving average, in some cases, is a bit more accurate because it takes into account scores that are a bit extreme or unique in the set (which in this case is the 10). The Moving Average tool also reveals a chart that plots the actual data against the averages for each point.

> Remember that the word *average* means only the one measure that best represents a set of scores, and that there are many different types of averages. Which type of average you use depends on the question that you are asking and the type of data you are trying to summarize. More about this later.

Computing a Weighted Mean

You've just seen an example of how to compute a simple mean. But there may be situations in which you have the occurrence of more than one value and you want to compute a weighted mean. A weighted mean can be computed easily by multiplying the value by the frequency of its occurrence, adding the total of all the products, and then dividing by the total number of occurrences.

To compute a weighted mean, follow these steps:

1. List all the values in the sample for which the mean is being computed, such as those shown in the column labeled Value (the value of X) in the following table.
2. List the frequency with which each value occurs.
3. Multiply the value by the frequency, as shown in the third column.
4. Sum all the values in the Value × Frequency column.
5. Divide by the total frequency.

For example, here's a table that organizes the values and frequencies in a flying proficiency test for 100 airline pilots.

Value	Frequency	Value × Frequency
97	4	388
94	11	1,034
92	12	1,104
91	21	1,911

Value	Frequency	Value × Frequency
90	30	2,700
89	12	1,068
78	9	702
60 (don't fly with this guy)	1	60
Total	100	8,967

The weighted mean is 8,967/100, or 89.67. Computing the mean this way is much easier than entering 100 different scores into your calculator or computer program.

How can we do this using Excel? Quite easily. Just follow these steps:

1. Enter the data you see in the above table in a new worksheet.

2. Create a formula like the one you see in Cell C2 to multiply the value times the frequency. You can see the formula in the formula bar in Figure 2.3.

3. Copy the formula down so that Cells C2 through C9 contain the multiplied values.

	A	B	C
1	Value	Frequency	Value x Frequency
2	97	4	=A2*B2
3	94	11	
4	92	12	
5	91	21	
6	90	30	
7	89	12	
8	78	9	
9	60	1	

Figure 2.3   Using Excel to Compute a Weighted Mean

More Excel

Remember that a cell contains information. Sometimes, that information is a formula, even though what you see is the results of that formula. In Figure 2.3, we used the Ctrl+` key combination to show the formulas or functions in a cell, rather than the results of that formula or function.

4. In Cells B10 and C10, use the SUM function to total the columns.

5. Now, placing the results in Cell C12, divide the total sum (in Cell C10) by the total frequency (in Cell B10), as you see in Figure 2.4. Ta-da! You did it again with a weighted average of 89.67, just as we showed above.

	A	B	C
1	Value	Frequency	Value x Frequency
2	97	4	388
3	94	11	1,034
4	92	12	1,104
5	91	21	1,911
6	90	30	2,700
7	89	12	1,068
8	78	9	702
9	60	1	60
10		100	8,967
11			
12		Weighted Average	89.67
13			

Figure 2.4   The Computation of a Weighted Mean

TECH TALK   In basic statistics, an important distinction needs to be made between those values associated with samples (a part of a population) and those associated with populations. To do this, statisticians use the following conventions. For a sample statistic (such as the mean of a sample), Roman letters are used. For a population parameter (such as the mean of a population), Greek letters are used. So, the mean for the spelling score for a sample of 100 fifth graders is represented by $\bar{X}_5$ whereas the mean for the spelling score for the entire population of fifth graders is represented as μ_5, using the Greek letter mu, or μ.

COMPUTING THE MEDIAN

The median is also an average, but of a very different kind. The **median** is defined as the **midpoint** in a set of scores. It's the point at which one half, or 50%, of the scores fall above and one half, or 50%, fall below. It's got some special qualities that we talk about later in this section, but for now, let's concentrate on how it is computed. There's no standard formula (but there is an Excel function, as we will see later) for computing the median.

To compute the median, follow these steps:

1. List the values in order, either from highest to lowest or lowest to highest.
2. Find the middle-most score. That's the median.

For example, here are the incomes from five different households:

$135,456

$25,500

$32,456

$54,365

$37,668

Here is the list ordered from highest to lowest:

$135,456

$54,365

$37,668

$32,456

$25,500

There are five values. The middle-most value is $37,668, and that's the median.

Now, what if the number of values is even? Let's add a value ($34,500) to the list so there are six income levels. Here they are:

$135,456

$54,365

$37,668

$34,500

$32,456

$25,500

When there is an even number of values, the median is simply the mean of two middle values. In this case, the middle two cases are $34,500 and $37,668. The mean of those two values is $36,084. That's the median for that set of six values.

What if the two middle-most values are the same, such as in the following set of data?

$45,678

$25,567

$25,567

$13,234

Then the median is the same as both of those middle-most values. In this case, it's $25,567.

And Now . . . Using Excel's MEDIAN Function

To compute the median of a set of numbers using Excel, follow these steps.

1. Enter the individual scores into one column in a worksheet, such as you see in Figure 2.5.

	A
1	Income Level
2	$135,456
3	$54,365
4	$37,668
5	$34,500
6	$32,456
7	$25,500

Figure 2.5   Data for Computing the Median

2. Select the cell into which you want to enter the **MEDIAN** function. In this example, we are going to compute the median in Cell A9.

3. Click on Cell A9 and type the median function as follows:

=MEDIAN(A2:A7)

and press the Enter key,

or

use the Insert → Function menu option and the "Inserting a Function" technique we talked about on pages 26 through 29 in Chapter 1 to enter the MEDIAN function.

You can see in Figure 2.6 the value of the median, and in the formula bar, you can see the MEDIAN function.

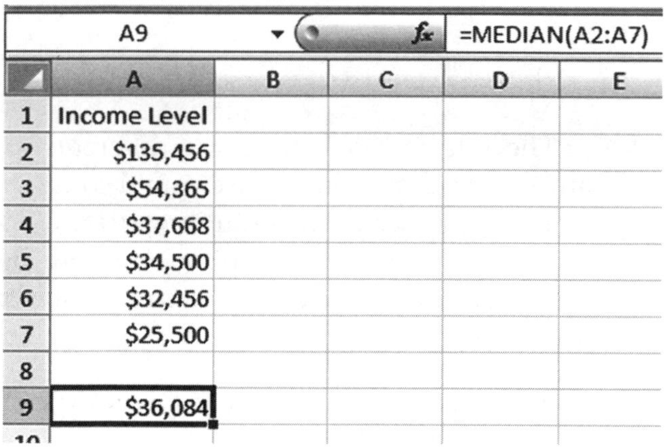

A9		f_x	=MEDIAN(A2:A7)	
A	**B**	**C**	**D**	**E**
1 Income Level				
2 $135,456				
3 $54,365				
4 $37,668				
5 $34,500				
6 $32,456				
7 $25,500				
8				
9 $36,084				
10				

Figure 2.6   Computing the Median Using the MEDIAN Function

TECH TALK   If you know about medians, you should know about **percentile points**. Percentile points are used to define the percentage of cases equal to and below a certain point in a distribution or set of scores. For example, if a score is "at the 75th percentile," it means that the score is at or above 75% of the other scores in the distribution. The median is also known as the 50th percentile, because it's the point below which 50% of the cases in the distribution fall. Other percentiles are useful as well, such as the 25th percentile, often called Q_1, and the 75th percentile, referred to as Q_3. So what's Q_2? The median, of course.

Here comes the answer to the question you've probably had in the back of your mind since we started talking about the median. Why use the median instead of the mean? For one very good reason. The median is insensitive to extreme scores, whereas the mean is not.

When you have a set of scores in which one or more scores are extreme, the median better represents the centermost value of that set of scores than any other measure of central tendency. Yes, even better than the mean.

What do we mean by extreme? It's probably easiest to think of an extreme score as one that is very different from the group to which it belongs. For example, consider the list of five incomes that we worked with earlier (shown again here):

$135,456

$54,365

$37,668

$32,456

$25,500

The value $135,456 is more different from the other four than any other value in the set. We would consider that an extreme score.

The best way to illustrate how useful the median is as a measure of central tendency is to compute both the mean and the median for a set of data that contains one or more extreme scores and then compare them to see which one best represents the group. Here goes.

The average or mean of the set of five scores you see above is the sum of the set of five divided by five, which turns out to be $57,089. On the other hand, the median for this set of five scores is $37,668. Which is more representative of the group? The value $37,668, because it clearly lies more in the middle of the group, and we like to think about the average as being representative or assuming a central position. In fact, the mean value of $57,089 falls above the fourth highest value ($54,365) and is not very central or representative of the distribution.

It's for this reason that certain social and economic indicators (mostly involving income) are reported using a median as a measure of central tendency, such as "The median income of the average American family is . . . ," rather than using the mean to summarize the values. There are just too many extreme scores that would **skew**, or significantly distort, what is actually a central point in the set or distribution of scores.

You learned earlier that sometimes the mean is represented by the capital letter M instead of $\bar{X}$. Well, other symbols are used for the median as well. We like the letter M, but some people confuse it with the mean, so they use Med for median, or Mdn. Don't let that throw you—just remember what the median is and what it represents, and you'll have no trouble adapting to different symbols.

More Excel

You can use the **QUARTILE** function to compute the 25th, 50th, and 75th percentiles as well as other quartiles in a distribution.

Remember the median? It's the second quartile in Excel's grand plan.

THINGS TO REMEMBER

Here are some interesting and important things to remember about the median.

- The mean is the middle point of a set of values, and the median is the middle point of a set of cases.
- Because the median cares about how many cases, and not the values of those cases, extreme scores (sometimes called **outliers**) don't count.

COMPUTING THE MODE

The third and last measure of central tendency that we'll cover, the **mode,** is the most general and least precise measure of central tendency, but it plays a very important part in understanding the characteristics of a special set of scores. The mode is the value that occurs most frequently. There is no formula for computing the mode.

To compute the mode, follow these steps:

1. List all the values in a distribution, but list each value only once.

2. Tally the number of times that each value occurs.

3. The value that occurs most often is the mode.

For example, an examination of the political party affiliation of 300 people might result in the following distribution of scores.

Party Affiliation	Number or Frequency
Democrats	90
Republicans	70
Independents	140

The mode is the value that occurs most frequently, which in the above example is Independents. That's the mode for this distribution.

And Now . . . Using *Excel's MODE Function*

To compute the mode of a set of numbers using Excel, follow these steps.

1. Enter the individual scores into one column in a worksheet as you see in Figure 2.7, where one column indicates party affiliation. Keep in mind that we have to enter a number (and not text), and Excel counts the number of times the value appears. We did include a code, though, in the same worksheet so you can keep straight what value stands for what party.

	A	B
1	Party	
2		1	1 = Democrats
3		2	2 = Republicans
4		3	3 = Independents
5		3	
6		3	
7		2	
8		2	
9		1	
10		1	
11		1	
12		1	
13		1	
14		2	
15		2	
16		2	
17		2	
18		1	
19		1	
20		1	

Figure 2.7   Data for Computing the Mode

2. Select the cell into which you want to enter the **MODE** function. In this example, we are going to compute the mode in Cell B21.

3. Click on Cell A5 and type the MODE function as follows:

=MODE(A2:A20)

and press the Enter key

or

Use the Insert → Function menu option and the "Inserting a Function" technique we talked about on pages 25 through 29 in Chapter 1 to enter the MODE function in Cell B21. You see the mode and the function in the formula bar in Figure 2.8.

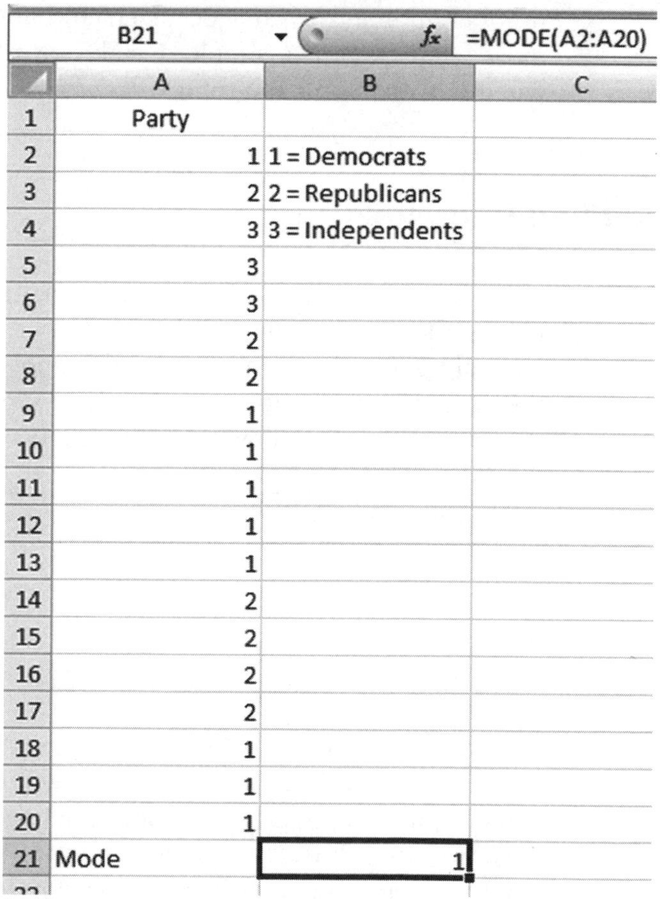

	B21	▼	f_x	=MODE(A2:A20)
	A	B		C
1	Party			
2		1	1 = Democrats	
3		2	2 = Republicans	
4		3	3 = Independents	
5		3		
6		3		
7		2		
8		2		
9		1		
10		1		
11		1		
12		1		
13		1		
14		2		
15		2		
16		2		
17		2		
18		1		
19		1		
20		1		
21	Mode		1	

Figure 2.8 Using the MODE Function

TECH
TALK You can use the COUNTIF function (in the category of functions under Database) to count the number of occurrences of text, which would be a really simple way to find out the mode without having to use the MODE function. Simply create a list, and then use the function defining the range and the values. In our example here, it would be something like =COUNTIF(A2:A20, Democrat), which would tally all the occurrences of the word Democrat and return it to the cell of your choice. Then, you would just select the largest value as the mode.

Want to know what the easiest and most commonly made mistake is when computing the mode? It's selecting the number of times a category occurs, rather than the label of the category itself.

Instead of the mode being Independents, it's easy for someone to conclude the mode is 140. Why? Because they are looking at the number of times the value occurred, and not the value that occurred most often! This is a simple mistake to make, so be on your toes when you are asked about these things.

Apple Pie à la Bimodal

If every value in a distribution contains the same number of occurrences, then there really isn't a mode (and Excel will return an incorrect value). But if more than one value appears with equal frequency, the distribution is multimodal. The set of scores can be bimodal (with two modes), as the following set of data using hair color illustrates.

Hair Color	Number or Frequency
Red	7
Blond	12
Black	45
Brown	45

In the above example, the distribution is bimodal because the frequency of the values of black and brown hair occurs equally. You can even have a bimodal distribution when the modes are relatively close together, but not exactly the same, such as 45 people with black hair and 44 with brown hair. The question becomes, How much does one class of occurrences stand apart from another? Can you have a trimodal distribution? Sure—where three values have the same frequency. It's unlikely, especially when you are dealing with a large set of data points, but certainly possible.

USING THE AMAZING ANALYSIS TOOLPAK TO COMPUTE DESCRIPTIVE STATISTICS

Now it's our first chance to use the amazing Analysis ToolPak that we introduced on page 37.

This particular item in the ToolPak, named Descriptive Statistics, computes more values than we need, but because you can't be selective in what the ToolPak computes (but can edit the results, as you will shortly see), we'll just show you all the results and deal only with those that we cover in this chapter. We'll follow the same procedure in later chapters.

To use the ToolPak to compute descriptive statistics, follow these steps. We're using the data you see in Figure 2.9, which you also saw way back in Figure 2.5.

	A
1	Income Level
2	$135,456
3	$54,365
4	$37,668
5	$34,500
6	$32,456
7	$25,500

Figure 2.9 Data for the Descriptive Statistics ToolPak Option

1. Click the Data tab → Data Analysis, and you will see the Data Analysis dialog box shown in Figure 2.10.

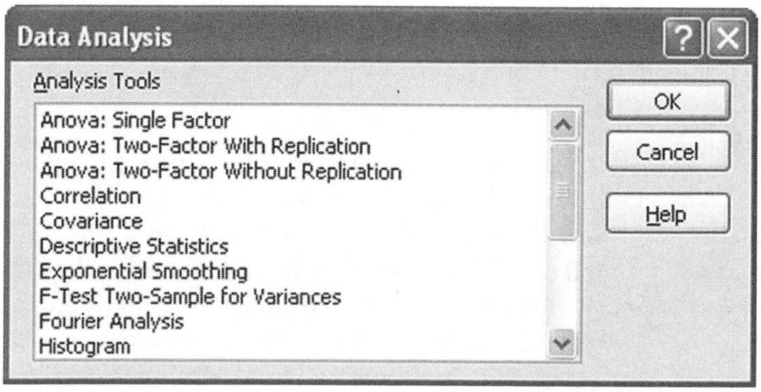

Data Analysis

Analysis Tools

- Anova: Single Factor
- Anova: Two-Factor With Replication
- Anova: Two-Factor Without Replication
- Correlation
- Covariance
- Descriptive Statistics
- Exponential Smoothing
- F-Test Two-Sample for Variances
- Fourier Analysis
- Histogram

OK Cancel Help

Figure 2.10 The Dialog Box That Gets Us Started With the Analysis ToolPak

2. Click Descriptive Statistics, and then click OK, and you will see the Descriptive Statistics dialog box as shown in Figure 2.11.

Figure 2.11   The Descriptive Statistics Dialog Box

3. Enter the range of data you want Excel to use in the computation of the descriptive statistics in the Input Range box, but also include the column heading (so the heading shows up in the analysis). In this example (as you can see in Figure 2.9), the data we want to analyze are in Cells A1 through A7.

4. Click the "Labels in First Row" check box.

5. Now click the Output Range button in the Output Options section of the dialog box, and enter the location where you want Excel to return the results of the analysis. In this example, we choose C1.

More Excel

Click or Drag?

This is covered in Appendix A, but it's important enough to emphasize again here. You can enter cell addresses in a dialog box in Excel in three ways.

You can just enter them using the keyboard (such as typing "A1").

You can click and drag the mouse over the cells you want to select, and then release the mouse and the cell range (or cell) will appear in the range box.

You can use the Collapse box (the one that looks like this: 🔳 . When you click this, it allows you to enter the range through dragging as well.

Which one for you? Whichever you find works best, but you have to know that typing cell ranges can get old very fast. Try to click and drag or use the Collapse box.

6. Click the "Summary statistics" check box in the Descriptive Statistics dialog box. The completed Descriptive Statistics dialog box is shown in Figure 2.12.

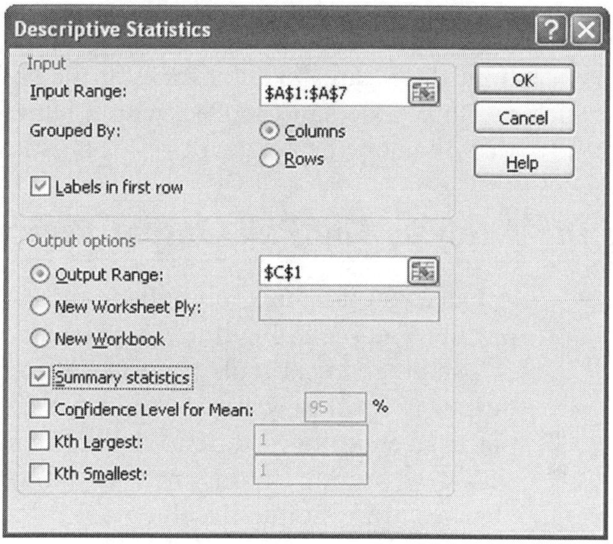

Figure 2.12 The Completed Descriptive Statistics Dialog Box

7. Click OK, and you will see the results, as shown in Figure 2.13.

	A	B	C	D
1	Income Level		*Income Level*	
2	$135,456			
3	$54,365		Mean	53324.17
4	$37,668		Standard I	16887.72
5	$34,500		Median	36084
6	$32,456		Mode	#N/A
7	$25,500		Standard I	41366.29
8			Sample Va	1.71E+09
9			Kurtosis	4.861219
10			Skewness	2.173756
11			Range	109956
12			Minimum	25500
13			Maximum	135456
14			Sum	319945
15			Count	6

Figure 2.13 The Descriptive Statistics ToolPak Results

Now this is pretty darn amazing. You get all this information with relatively few clicks. You just have to make sure that you get your ducks all lined up in a row (where that expression ever came from, who knows?), but you must be sure that all the cell references are entered accurately.

You can see all kinds of useful information in Figure 2.13 ranging from the Mean (53324.16667) of the six values (See how Count = 6?) to the Median (36084), with a bunch of other stuff (some of it we will not be dealing with—see Statistics 2 in your course catalog).

Make the Analysis ToolPak Output Pretty

Once you use any tool in the Analysis ToolPak and get some output like you see in Figure 2.13, you can (of course) leave it like it is or use other Excel tools to format it to better fit your needs. This output is absolutely part of the worksheet you created in the first place, so anything you do to the entire sheet also has an impact on this new output. In Figure 2.14, you can see we made several changes using simple Excel tools.

	A	B	C	D
1	Income Level		Income Level	
2	$135,456			
3	$54,365		Mean	53,324.17
4	$37,668		Standard Error	16,887.72
5	$34,500		Median	36,084.00
6	$32,456		Standard Deviation	41,366.29
7	$25,500		Sample Variance	1,711,170,163.37
8			Kurtosis	4.86
9			Skewness	2.17
10			Range	109,956.00
11			Minimum	25,500.00
12			Maximum	135,456.00
13			Sum	319,945.00
14			Count	6.00
15				

Figure 2.14   The New and Improved Descriptive Statistics Output

- We formatted the entire worksheet in Arial 12.
- We formatted numbers so they made more sense. For example, we reformatted the mean as $53,324.17 from 53324.16667.
- We deleted the Mode cells because there is no mode.
- We used the Format → Column → AutoFit option to adjust the columns so that all the information fit on the worksheet.

- We could have added other things and used more of Excel's bells and whistles (color, table formats, shading, etc.), but what you see does a fine job of showing the results of the analysis. Fancy is nice, but there's nothing wrong with simple and straightforward—words to live by.

WHEN TO USE WHAT

OK, we've defined three different measures of central tendency and given you fairly clear examples of each. But the most important question remains unanswered. That is, "When do you use which measure?"

In general, which measure of central tendency you use depends on the type of data that you are describing. Unquestionably, a measure of central tendency for qualitative, categorical, or nominal data (such as racial group, eye color, income bracket, voting preference, and neighborhood location) can be described using only the mode.

For example, you can't be looking at the most central measure that describes which political affiliation is most predominant in a group and use the mean—what in the world could you conclude, that everyone is half Republican? Rather, that out of 300 people, almost half (140) are Independent seems to be the best way of describing the value of this variable. In general, the median and mean are best used with quantitative data, such as height, income level in dollars (not categories), age, test score, reaction, and number of hours completed for a degree.

It's also fair to say that the mean is a more precise measure than the median, and the median is a less precise measure than the mode. This means that all other things being equal, use the mean, and indeed, the mean is the most often used measure of central tendency. However, we do have occasions when the mean would not be appropriate as a measure of central tendency—for example, when we have categorical or nominal data, such as hair color. Then we use the mode. So, here is a set of three guidelines that may be of some help. And remember, there can always be exceptions.

1. Use the mode when the data are categorical in nature and values can fit into only one class, such as hair color, political affiliation, neighborhood location, and religion. When this is the case, these categories are called mutually exclusive.

2. Use the median when you have extreme scores and you don't want to distort the average, such as income.

3. Finally, use the mean when you have data that do not include extreme scores and are not categorical, such as the numerical score on a test or the number of seconds it takes to swim 50 yards.

Summary

No matter how fancy schmancy your statistical techniques are, you will still almost always start by simply describing what's there—hence, the importance of understanding the simple notion of central tendency. From here, we go to another important descriptive construct: variability, or how different scores are from one another.

Time to Practice

1. Compute the mean, median, and mode for the following three sets of scores saved as Chapter 2 Data Set 1. Do it by hand or using Excel. Show your work, and if you using Excel, print out a copy of the output.

Score 1	Score 2	Score 3
3	34	154
7	54	167
5	17	132
4	26	145
5	34	154
6	25	145
7	14	113
8	24	156
6	25	154
5	23	123

2. You are the manager of a fast-food store. Part of your job is to report to the boss at the end of each day what special is selling best. Use your vast knowledge of descriptive statistics and write one paragraph to let the boss know what happened today. Here are the data. Do this exercise by hand. Be sure to include a copy of your work.

Special	Number Sold	Cost
Huge Burger	20	$2.95
Baby Burger	18	$1.49
Chicken Littles	25	$3.50
Porker Burger	19	$2.95
Yummy Burger	17	$1.99
Coney Dog	20	$1.99
Total Specials Sold	119	

3. Under what conditions would you use the median rather than the mean as a measure of central tendency? Why? Provide an example of two situations in which the median might be more useful than the mean as a measure of central tendency.

4. You're in business for yourself and you have been fortunate to buy and own the Web site titled havefun.com, where you sell every imaginable stupid toy (like potato guns) and game that everyone needs. You're reviewing your advertising budget for the third quarter (from July 1 through September 31) and need to compute the mean (an average, remember?). Here are the sales data in dollars. Use Excel to compute the average sales by toy and by month.

Toy	July Sales	August Sales	September Sales	Average Sale
Slammer	12,345	14,453	15,435	
Radar Zinger	31,454	34,567	29,678	
LazerTags	3,253	3,121	5,131	
Average Sale				

5. As the head of public health, you do a weekly census across age groups of the number of cases of flu reported. By hand, compute the mean and the median by week. Which do you think, given this particular data, is the most useful measure of central tendency?

	12/1 Through 12/7	12/8 Through 12/15	12/16 Through 12/23
0–4 years	12	14	15
5–9 years	15	12	14
10–14 years	12	24	21
15–19 years	38	12	19
Mean			
Median			

Answers to Practice Questions

1. By hand . . .

	Score 1	Score 2	Score 3
Mean	5.6	27.6	144.3
Median	5.5	25.0	149.5
Mode	5	25.34	154

Figure 2.15 shows you what the Excel worksheet could look like. Keep in mind that there are two modes in the set of data named Score 2 (they are 34 and 25), even though Excel will show only one, which in this case is 34.

	A	B	C	D
1		Score 1	Score 2	Score 3
2		3	34	154
3		7	54	167
4		5	17	132
5		4	26	145
6		5	34	154
7		6	25	145
8		7	14	113
9		8	24	156
10		6	25	154
11		5	23	123
12	Mean	5.6	27.6	144.3
13	Median	5.5	25	149.5
14	Mode	5	34	154
15				

Figure 2.15   Using Excel to Compute the Mean, Median, and Mode for the Data From Question 1

2. Here's what your one paragraph might look like.

   As usual, the Chicken Littles [the mode] led the way in sales. The total amount of food sold was $303, for an average of $2.55 for each special.

3. You use the median when you have extreme scores, which would disproportionately bias the mean. One situation where the median is preferable to the mean is where income is reported. Because it varies so much, you want a measure of central tendency that is insensitive to extreme scores. Another example is where you have an extreme score or an outlier, such as the speed with which a group of adolescents can run 100 yards, where there are one or two exceptionally fast (or slow) individuals.

4. Averages for the quarter by month are shown in Figure 2.16. We inserted the AVERAGE function in only one cell and then copied it across rows and only one other cell and copied it across columns.

	A	B	C	D	E
1	Toy	July Sales	August Sales	September Sales	Average Sales
2	Slammer	$12,345.00	$ 14,453.00	$ 15,435.00	$ 14,077.67
3	Radar Zinger	$31,454.00	$ 34,567.00	$ 29,678.00	$ 31,899.67
4	Potato Gun	$ 3,253.00	$ 3,121.00	$ 5,131.00	$ 3,835.00
5	Average Sale	$15,684.00	$ 17,380.33	$ 16,748.00	$ 16,604.11
6					

Figure 2.16   Average Sales

5.

	12/1 Through 12/7	12/8 Through 12/15	12/16 Through 12/23
Mean	19.25	15.5	17.25
Median	13.5	13	17

At least for the week of 12/1 through 12/7, the median is a better representative measure of central tendency because there is one extreme data point, 38, for the 15- to 19-year-old group. For all the other weeks, the mean seems appropriate.

3 Vive la Différence

Understanding Variability

Difficulty Scale ☺☺☺☺ (moderately easy, but not a cinch)

How much Excel? ⬚ ⬚ ⬚ ⬚ ⬚ (a ton)

What you'll learn about in this chapter

- Why variability is valuable as a descriptive tool
- How to compute the range, standard deviation, and variance
- How the standard deviation and variance are alike, and how they are different
- Using the Analysis ToolPak to compute the range, standard deviation, and variance

WHY UNDERSTANDING VARIABILITY IS IMPORTANT

In Chapter 2, you learned about different types of averages, what they mean, how they are computed, and when to use them. But when it comes to descriptive statistics and describing the characteristics of a distribution, averages are only half the story. The other half is measures of variability.

In the most simple of terms, **variability** reflects how scores differ from one another. For example, the following set of scores shows some variability:

7, 6, 3, 3, 1

The following set of scores has the same mean (4) and has less variability than the previous set:

3, 4, 4, 5, 4

The next set has no variability at all—the scores do not differ from one another—but it also has the same mean as the other two sets we just showed you.

4, 4, 4, 4, 4

Variability (also called spread or dispersion) can be thought of as a measure of how different scores are from one another. It's even more accurate (and maybe even easier) to think of variability as how different scores are from one particular score. And what "score" do you think that might be? Well, instead of comparing each score to every other score in a distribution, the one score that could be used as a comparison is—that's right—the mean. So, variability becomes a measure of how much each score in a group of scores differs from the mean. More about this in a moment.

Remember what you already know about computing averages— that an average (whether it is the mean, the median, or the mode) is a representation of a set of scores. Now, add your new knowledge about variability—that it reflects how different scores are from one another. Each is an important descriptive statistic. Together, these two (average and variability) can be used to describe the characteristics of a distribution and show how distributions differ from one another.

Three measures of variability are commonly used to reflect the degree of variability, spread, or dispersion in a group of scores.

These are the range, the standard deviation, and the variance. Let's take a closer look at each one and how each one is used.

COMPUTING THE RANGE

The range is the most general measure of variability. It gives you an idea of how far apart scores are from one another. The **range** is computed simply by subtracting the lowest score in a distribution from the highest score in the distribution.

In general, the formula for the range is

$$r = h - l,$$ (3.1)

where

    r   is the range,

    h   is the highest score in the data set, and

    l   is the lowest score in the data set.

Take the following set of scores, for example (shown here in descending order):

$$98, 86, 77, 56, 48$$

In this example, 98 − 48 = 50. The range is 50.

TECH TALK   There really are two kinds of ranges. One is the exclusive range, which is the highest score minus the lowest score (or $h - l$) and the one we just defined. The second kind of range is the inclusive range, which is the highest score minus the lowest score plus 1 (or $h - l + 1$). You most commonly see the exclusive range in research articles, but the inclusive range is also used on occasion if the researcher prefers it.

The range is used almost exclusively to get a very general estimate of how wide or different scores are from one another—that is, the range shows how much spread there is from the lowest to the highest point in a distribution.

So, although the range is fine as a general indicator of variability, it should not be used to reach any conclusions regarding how individual scores differ from one another. Remember, the range uses only two scores—a poor reflection of what's really happening.

And as far as Excel is concerned, there is no function for computing the range. Rather, you can create a simple formula that subtracts one value from the other and adds 1 (to compute an inclusive range) or doesn't add anything for the exclusive range.

COMPUTING THE STANDARD DEVIATION

Now we get to the most frequently used measure of variability, the standard deviation. Just think about what the term implies; it's a deviation from something (guess what?) that is standard. Actually, the **standard deviation** (abbreviated as *s* or *SD*) represents the average amount of variability in a set of scores. In practical terms, it's the average distance from the mean. The larger the standard deviation, the larger the average distance each data point is from the mean of the distribution.

So, what's the logic behind computing the standard deviation? Your initial thoughts may be to compute the mean of a set of scores and then subtract each individual score from the mean. Then, compute the average of that distance. That's a good idea—you'll end up with the average distance of each score from the mean. But it won't work (see if you know why even though we'll show you why in a moment).

First, here's the formula for computing the standard deviation:

$$s = \sqrt{\frac{\sum (X - \bar{X})^2}{n - 1}}, \tag{3.2}$$

where

s is the standard deviation;

$\sum$ is sigma, which tells you to find the sum of what follows;

X is each individual score;

$\bar{X}$ is the mean of all the scores; and

n is the sample size.

This formula finds the difference between each individual score and the mean $(X - \bar{X})$, squares each difference, and sums them all together. Then, it divides the sum by the size of the sample (minus 1) and takes the square root of the result. As you can see, and as we mentioned earlier, the standard deviation is an average deviation from the mean.

Here are the data we'll use in the following step-by-step explanation of how to compute the standard deviation.

$$5, 8, 5, 4, 6, 7, 8, 8, 3, 6$$

1. List each score. It doesn't matter whether the scores are in any particular order.

2. Compute the mean of the group.

3. Subtract the mean from each score.

Here's what we've done so far, where $X - \bar{X}$ represents the difference between the actual score and the mean of all the scores, which is 6.

X	$\overline{X}$	$X - \overline{X}$
8	6	8 − 6 = +2
8	6	8 − 6 = +2
8	6	8 − 6 = +2
7	6	7 − 6 = +1
6	6	6 − 6 = 0
6	6	6 − 6 = 0
5	6	5 − 6 = −1
5	6	5 − 6 = −1
4	6	4 − 6 = −2
3	6	3 − 6 = −3

4. Square each individual difference. The result is the column marked $(X - \overline{X})^2$.

X	$(X - \overline{X})$	$(X - \overline{X})^2$
8	+2	4
8	+2	4
8	+2	4
7	+1	1
6	0	0
6	0	0
5	−1	1
5	−1	1
4	−2	4
3	−3	9
Sum	0	28

5. Sum all the squared deviations about the mean. As you can see above, the total is 28.

6. Divide the sum by $n - 1$, or $10 - 1 = 9$, so then $28/9 = 3.11$.

7. Compute the square root of 3.11, which is 1.76 (after rounding). That is the standard deviation for this set of 10 scores.

And Now . . . Using Excel's STDEV Function

To compute the standard deviation of a set of numbers using Excel, follow these steps:

1. Enter the individual scores into one column in a worksheet such as you see in Figure 3.1.

	A	B
1		Score
2		8
3		8
4		8
5		7
6		6
7		6
8		5
9		5
10		4
11		3
12		

Figure 3.1   Data for the STDEV Function

2. Select the cell into which you want to enter the **STDEV** function. In this example, we are going to compute the standard deviation in Cell B12.

3. Now click on Cell B12 and type the STDEV function as follows:
   =STDEV(B2:B11)
   and press the Enter key,
   or
   use the Formulas → Insert Function menu option and the "Inserting a Function" technique we talked about on pages 25 through 29 in Chapter 1 to enter the STDEV function in Cell B12.

4. As you can see in Figure 3.2, the standard deviation was computed and the value (1.76, just as we did manually earlier) returned to Cell B12. Notice that in the formula bar in Figure 3.2, you can see the STDEV function fully expressed.

B12	▼		f_x	=STDEV(B2:B11)	
	A	B	C	D	E
1		Score			
2		8			
3		8			
4		8			
5		7			
6		6			
7		6			
8		5			
9		5			
10		4			
11		3			
12		1.76			

Figure 3.2   The Computation of the Standard Deviation Using the STDEV Function

More Excel

What Function Loves Ya, Baby

Now, here's a surprise. When you go to select the STDEV function from the Formulas → Insert Function options, you see that there are actually several functions that can compute the standard deviation. There are two especially important ones for us. One is named STDEV (the one we used in the example above), and the second is named STDEVP. They both compute the standard deviation of a set of scores, but what's the difference?

The one that we used, STDEV, computes the standard deviation for a set of scores from a sample. STDEVP computes the standard deviation for a set of scores that is an entire population. What's the difference in the two values, and why use one rather than the other? More about this later, but for now, because STDEV is based on only a portion of the entire population (the definition of a sample, right?), it's more likely (we're talking probability here) to overestimate the true, real, honest-to-Pete value of the standard deviation when computed for the entire population.

When to use what? Use STDEV—you'll almost always be correct. But if a group of scores is defined as a population, then STDEVP is the function that loves ya, baby.

What we now know from these results is that each score in this distribution differs from the mean by an average of 1.76 points.

Let's take a short step back and examine some of the operations in the standard deviation formula. They're important to review and will increase your understanding of what the standard deviation is.

First, why didn't we just add up the deviations from the mean? Because the sum of the deviations from the mean is always equal to 0. Try it by summing the deviations $(2 + 2 + 2 + 1 + 0 + 0 - 1 - 1 - 2 - 3)$. In fact, that's the best way to check if you computed the mean correctly.

TECH TALK

There's another type of deviation that you may read about, and you should know what it means. The **mean deviation** (also called the mean absolute deviation) is the average of the absolute value of the deviations from the mean. You already know that the sum of the deviations from the mean must equal 0 (otherwise, the mean probably has been computed incorrectly). Instead, let's take the absolute value of each deviation (which is the value regardless of the sign). Sum them together and divide by the number of data points, and you have the mean deviation. (Note: The absolute value of a number is usually represented as that number with a vertical line on each side of it, such as |5|. For example, the absolute value of –6, or |–6|, is 6.)

Second, why do we square the deviations? Because we want to get rid of the negative sign so that when we do eventually sum them, they don't add up to 0.

And finally, why do we eventually end up taking the square root of the entire value in Step 7? Because we want to return to the same units with which we originally started. We squared the deviations from the mean in Step 4 (to get rid of negative values) and then took the square root of their total in Step 7. Pretty tidy.

Why n – 1? What's Wrong With Just n?

You might have guessed why we square the deviations about the mean and why we go back and take the square root of their sum. But how about subtracting the value of 1 from the denominator of the formula? Why do we divide by $n - 1$ rather than just plain ol' n? Good question.

The answer is that s (the standard deviation) is an estimate of the population standard deviation and is an **unbiased estimate** at that, but only when we subtract 1 from n. By subtracting 1 from the denominator, we artificially force the standard deviation to be larger than it would be otherwise. Why would we want to do that? Because, as good scientists, we are conservative, and that is exactly what we are doing by using STDEV rather than the STDEVP function in the earlier example.

In fact, you can see that if you look at Figure 3.3, where we used both functions to compute the standard deviation. See how the unbiased one (1.76) is larger than the biased one (1.67)? That's because the one based on the sample (the unbiased one) intentionally overestimates the value.

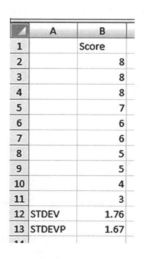

	A	B
1		Score
2		8
3		8
4		8
5		7
6		6
7		6
8		5
9		5
10		4
11		3
12	STDEV	1.76
13	STDEVP	1.67

Figure 3.3 A Comparison of the STDEV and STDEVP Functions

Being conservative means that if we have to err, we will do so on the side of overestimating what the standard deviation of the population is. Dividing by a smaller denominator lets us do so. Thus, instead of dividing by 10, we divide by 9. Or, instead of dividing by 100, we divide by 99.

 TECH TALK Biased estimates are appropriate if your intent is only to describe the characteristics of the sample. But if you intend to use the sample as an estimate of a population parameter, then the unbiased statistic is best to calculate.

Take a look in the following table and see what happens as the size of the sample gets larger (and moves closer to the population in size). The $n - 1$ adjustment has far less of an impact on the difference between the biased and the unbiased estimates of the standard deviation (the bold column in the table). All other things being equal, then, the larger the size of the sample, the less of a difference there is between the biased and the unbiased estimates of the standard deviation. Check out the following table, and you'll see what we mean.

Sample Size	Value of Numerator in Standard Deviation	Biased Estimate of the Population Standard Deviation (dividing by n)	Unbiased Estimate of the Population Standard Deviation (dividing by n – 1)	Difference Between Biased and Unbiased Estimates
10	500	7.07	7.45	**.38**
100	500	2.24	2.25	**.01**
1,000	500	0.7071	0.7075	**.0004**

The moral of the story? When you compute the standard deviation for a sample, which is an estimate of the population, the closer to the size of the population the sample is, the more accurate the estimate will be.

What's the Big Deal?

The computation of the standard deviation is very straightforward. But what does it mean? As a measure of variability, all it tells us is how much each score in a set of scores, on the average, varies from the mean. But it has some very practical applications, as you will find out in Chapter 4. Just to whet your appetite, consider this:

The standard deviation can be used to help us compare scores from different distributions, *even when the means and standard deviations are different.*

Amazing! This, as you will see, can be very cool.

THINGS TO REMEMBER

The standard deviation is computed as the average distance from the mean. So, you will need to first compute the mean as a measure of central tendency. Don't fool around with the median or the mode in trying to compute the standard deviation. The larger the standard deviation, the more spread out the values are, and the more different they are from one another.

Just like the mean, the standard deviation is sensitive to extreme scores. When you are computing the standard deviation of a sample and you have extreme scores, note that somewhere in your written report.

If $s = 0$, there is absolutely no variability in the set of scores, and they are essentially identical in value. This will rarely happen.

COMPUTING THE VARIANCE

Here comes another measure of variability and a nice surprise. If you know the standard deviation of a set of scores and you can square a number, you can easily compute the variance of that same set of scores. This third measure of variability, the **variance,** is simply the standard deviation squared. In other words, it's the same formula you saw earlier, without the square root bracket, like the one shown in Formula 3.3:

$$s^2 = \frac{\sum (X - \bar{X})^2}{n - 1}. \tag{3.3}$$

If you take the standard deviation and never complete the last step (taking the square root), you have the variance. In other words, $s^2 = s \times s$, or the variance equals the standard deviation times itself (or squared). In our earlier example, where the standard deviation was equal to 1.76, the variance is equal to 1.76^2 or 3.10.

And Now . . . Using Excel's VAR Function

To compute the variance of a set of numbers using Excel, follow these steps:

1. Enter the individual scores into one column in a worksheet. We're going to use the same scores as those shown in Figure 3.1.

2. Select the cell into which you want to enter the **VAR** function. In this example, we are going to compute the variance in Cell B12.

3. Now click on Cell B12 and type the VAR function as follows:
   =VAR(B2:B11)
   and press the Enter key,
   or
   use the Formula → Insert Function menu option and the "Inserting a Function" technique we talked about on pages 25 through 29 in Chapter 1 to enter the VAR function in Cell B12.

4. As you can see in Figure 3.4, the variance was computed and the value returned to Cell B12. Notice that in the formula bar in Figure 3.4, you can see the VAR function fully expressed and the value computed as 3.10, just as we did manually earlier in the chapter. You will notice that we computed this value to be 3.10, but Figure 3.4 shows it to be 3.11. That's a function of the internal formula that Excel uses to compute the variance and nothing more.

B12			f_x	=VAR(B2:B11)	
	A	B	C	D	E
1		Score			
2		8			
3		8			
4		8			
5		7			
6		6			
7		6			
8		5			
9		5			
10		4			
11		3			
12		3.11			

Figure 3.4   Using the VAR Function to Compute the Variance

More Excel

STDEV Is to STDEVP as VAR Is to VARP

Sounds like an acronym love fest, right? Nope—just as there is a corresponding function for computing the standard deviation of a population (STDEVP), so there is for the variance as well, and it is named **VARP.** In Figure 3.5, you can see how when each is computed, like the unbiased nature of STDEV or VAR, they are larger than their STDEVP and VARP cousins.

	A	B
1		Score
2		8
3		8
4		8
5		7
6		6
7		6
8		5
9		5
10		4
11		3
12	VAR	3.11
13	VARP	2.80

Figure 3.5   Comparing the VAR and VARP Functions

You are not likely to see the variance mentioned by itself in a journal article or see it used as a descriptive statistic. This is because the variance is a difficult number to interpret and to apply to a set of data. After all, it is based on squared deviation scores.

But the variance is important because it is used both as a concept and as a practical measure of variability in many statistical formulas and techniques. You will learn about these later in *Statistics for People Who (Think They) Hate Statistics . . . Excel 2007 Edition.*

The Standard Deviation Versus the Variance

How are standard deviation and the variance the same, and how are they different?

Well, they are both measures of variability, dispersion, or spread. The formulas used to compute them are very similar. You see them all over the place in the "Results" sections of journals.

They are also quite different.

First, and most important, the standard deviation (because we take the square root of the average summed squared deviation) is stated in the original units from which it was derived. The variance is stated in units that are squared (the square root is never taken).

What does this mean? Let's say that we need to know the variability of a group of production workers assembling circuit boards. Let's say that they average 8.6 boards per hour, and the standard deviation is 1.59. The value 1.59 means that the difference in the average number of boards assembled per hour is about 1.59 circuit boards from the mean.

Let's look at an interpretation of the variance, which is 1.59^2, or 2.53. This would be interpreted as meaning that the average difference between the workers is about 2.53 circuit boards *squared* from the mean. Which of these two makes more sense?

USING THE AMAZING ANALYSIS TOOLPAK (AGAIN!)

Guess what? We already did this! See pages 57 and 58 in Chapter 2 where the ToolPak does a descriptive analysis including the standard deviation, the variance, and the range. Now, that was easy, wasn't it?

Summary

Measures of variability help us understand even more fully what a distribution of data points looks like. Along with a measure of central tendency, we can use these values to distinguish distributions from one another and effectively describe what a collection of test scores, heights, or measures of personality looks like. Now that we can think and talk about distributions, let's look at ways we can look at them.

Time to Practice

1. Why is the range the most convenient measure of dispersion, yet the most imprecise measure of variability? When would you use the range?

2. For the following set of scores, compute the range, the unbiased and the biased standard deviation, and the variance. Do the exercise by hand.

    31, 42, 35, 55, 54, 34, 25, 44, 35

3. Why is the unbiased estimate a larger value than the biased estimate?

4. This practice problem uses the data contained in the file named Chapter 3 Data Set 1 in Appendix C. There are two variables in this data set.

Variable	Definition
Height	height in inches
Weight	weight in pounds

Using Excel, compute all of the measures of variability you can for height and weight. Does Excel compute a biased or an unbiased estimate? How do you know?

5. You're one of the outstanding young strategists for Western Airlines and are really curious to examine the number of passengers flying in a morning–evening flight comparison out of the Kansas City; Philadelphia; and Providence, Rhode Island, hubs for the last 2 days of last week. Here are the data—compute the descriptive statistics you need that you will use in tomorrow's presentation to your boss and provide a few sentences of summary. Be on time.

	Thursday	Friday	Thursday	Friday	Thursday	Friday
Morning Flights	To Kansas City	To Kansas City	To Philadelphia	To Philadelphia	To Providence	To Providence
Number of passengers	258	251	303	312	166	176
Evening Flights	To Kansas City	To Kansas City	To Philadelphia	To Philadelphia	To Providence	To Providence
Number of passengers	312	331	321	331	210	274

Answers to Practice Questions

1. The range is the most convenient measure of dispersion because it requires you only to subtract one number (the lowest value) from another number (the highest value). It's imprecise because it does not take into account the values that fall between the highest and the lowest values in a distribution. Use the range when you want a very gross (and not very precise) estimate of the variability in a distribution.

2. The range is 30. The unbiased sample standard deviation equals 10.19. The biased estimate equals 9.60. The difference is due to dividing by a sample size of 8 (for the unbiased estimate) as compared to a sample size of 9 (for the biased estimate). The unbiased estimate of the variance is 103.78, and the biased estimate is 92.25.

3. The unbiased estimate should be larger than the biased estimate because we are intentionally being conservative and overestimating the size of the population standard deviation. Arithmetically, for the biased estimate, the denominator is larger and hence the value is smaller.

4. Figure 3.6 shows you the data and the computed standard deviation and variance. In both cases, we used the unbiased functions STDEV and VAR to compute the values. The only way to know whether the

functions compute a biased or an unbiased estimate is if you know which function was selected to compute these values. After the fact, you can, of course, just look at the formula bar and examine the cell's contents.

	A	B	C
1		Height	Weight
2		53	156
3		46	131
4		54	123
5		44	142
6		56	156
7		76	171
8		87	143
9		65	135
10		45	138
11		44	114
12		57	154
13		68	166
14		65	153
15		66	140
16		54	143
17		66	156
18		51	173
19		58	143
20		49	161
21		48	131
22	Range	43	59
23	Standard Deviation	11.44	15.65
24	Variance	130.78	245.00
25			

Figure 3.6   Computing the Standard Deviation and the Variance Using STDEV and VAR

5. First, the facts . . . The average number of folks flying on a morning flight is 244 (actually 244.33), and the average number flying on an evening flight is 296 (actually 296.5).

The standard deviation for morning flights is 61.74 and for evening flights is 47.35.

You can compute just about any measures of central tendency and variation and make some sense of them. For example,

Just on the surface (and there are a lot more descriptive things we could compute that might be of interest, such as a city by time-of-day comparison), more people fly in the evening than the morning (which might mean more one-way tickets are purchased, etc.) and the number of people tends to be more consistent (a lower standard deviation) when it comes to evening flights.

A Picture Really Is Worth a Thousand Words

Difficulty Scale ☺☺☺☺ (pretty easy, but not a cinch)

How much Excel? 📊 📊 📊 📊 (lots and lots)

What you'll learn about in this chapter

- Why a picture is really worth a thousand words
- How to create a histogram and a polygon
- Using the Analysis ToolPak to create a histogram
- Using the SKEW and KURT functions
- Using Excel to create charts
- Using Excel to modify charts
- Different types of charts and their uses
- What pivot tables are and how to use them

WHY ILLUSTRATE DATA?

In the previous two chapters, you learned about two important types of descriptive statistics—measures of central tendency and measures of variability. Both of these provide you with the one best score for describing a group of data (central tendency) and a measure of how diverse, or different, scores are from one another (variability).

What we did not do, and what we will do here, is examine how differences in these two measures result in different-looking distributions. Numbers alone (such as $\bar{X} = 10$ and $s = 3$) may be important, but a visual representation is a much more effective way of examining the characteristics of a distribution as well as the characteristics of any set of data.

81

So, in this chapter, we'll learn how to visually represent a distribution of scores as well as how to use different types of graphs to represent different types of data. And, at the end, we'll introduce pivot tables, a relatively new addition to Excel's feature that allows you to manipulate rows and columns in tables to appear as you want.

TEN WAYS TO A GREAT FIGURE (EAT LESS AND EXERCISE MORE?)

Whether you create illustrations by hand or use a computer program, the principles of decent design still apply. Here are 10 to copy and put above your desk.

1. *Minimize chart or graph junk.* "Chart junk" (a close cousin to "word junk") is where you use every function, every graph, and every feature a computer program has to make your charts busy, full, and uninformative. Less is definitely more.

2. *Plan out your chart before you start creating the final copy.* Use graph paper to create a preliminary sketch even if you will be using a computer program to generate the graph.

3. *Say what you mean and mean what you say—no more and no less.* There's nothing worse than a cluttered (with too much text and fancy features) graph to confuse the reader. And, there's nothing better than a graph that is simple and straightforward.

4. *Label everything so nothing is left to the misunderstanding of the audience.*

5. *A graph should communicate only one idea.*

6. *Keep things balanced.* When you construct a graph, center titles and axis labels.

7. *Maintain the scale in a graph.* The scale refers to the relationship between the horizontal and vertical axes. This ratio should approximate the "golden triangle," which is about 3:4, so a graph that is 3 inches tall will be about 4 inches wide.

8. *Simple is best.* Keep the chart simple, but not simplistic. Convey the one idea as straightforwardly as possible, with distracting information saved for the accompanying text.

Remember, a chart or graph should be able to stand alone, and the reader should be able to understand the message.

9. *Limit the number of words you use.* Too many words, or words that are too large, can detract from the visual message your chart should convey.

10. *A chart alone should convey what you want to say.* If it doesn't, go back to your plan and try it again.

Want to read the best resource in the universe on how to make data visually attractive and informative? Get any of Edward Tufte's self-published and hugely successful books, such as *The Visual Display of Quantitative Information* from Graphics Press. This book has become a classic for explaining how numerical data can be illustrated.

FIRST THINGS FIRST: CREATING A FREQUENCY DISTRIBUTION

The most basic way to illustrate data is through the creation of a frequency distribution. A **frequency distribution** is a method of tallying, and representing, how often certain scores occur. In the creation of a frequency distribution, scores are usually grouped into **class intervals,** or ranges of numbers.

Here are 50 scores on a test of reading comprehension and what the frequency distribution for these scores looks like. Here are the raw data on which the distribution is based.

47	10	31	25	20
2	11	31	25	21
44	14	15	26	21
41	14	16	26	21
7	30	17	27	24
6	30	16	29	24
35	32	15	29	23
38	33	19	28	20
35	34	18	29	21
36	32	16	27	20

And here's the frequency distribution.

Class Interval	Frequency
45–49	1
40–44	2
35–39	4
30–34	8
25–29	10
20–24	10
15–19	8
10–14	4
5–9	2
0–4	1

The Classiest of Intervals

As you can see from the above table, a class interval is a range of numbers, and the first step in the creation of a frequency distribution is to define how large each interval will be. You can see in the frequency distribution that we created, each interval spans five possible scores such as 5–9 (which contains scores 5, 6, 7, 8, and 9) and 40–44 (which contains scores 40, 41, 42, 43, and 44).

How did we decide to have an interval that contains only five scores? Why not five intervals each consisting of 10 scores? Or two each consisting of 25 scores? Here are some general rules to follow in the creation of a class interval, regardless of the size of values in the data set you are using.

1. Select a class interval that has a range of 2, 5, 10, or 20 data points. In our example, we chose 5.

2. Select a class interval so that 10 to 20 such intervals cover the entire set of data. A convenient way to do this is to compute the range, then divide by a number that represents the number of intervals you want to use (between 10 and 20). In our example, there are 50 scores and we wanted 10 intervals: 50/10 = 5, which is the size of each class interval. If you had a set of scores ranging from 100 to 400, you can start with the following estimate and work from there: 300/20 = 15, so 15 would be the class interval.

3. Begin listing the class interval with a multiple of that interval. In our frequency distribution shown earlier, the class interval is 5, and we started with the lowest class interval of 0.

4. Finally, the largest interval goes at the top of the frequency distribution.

Once class intervals are created, it's time to complete the frequency part of the frequency distribution. That's simply counting the number of times a score occurs in the raw data and entering that number in each of the class intervals represented by the count.

In the frequency distribution that we created earlier (on page 84), the number of scores that occur between 30 and 34 and are in the 30–34 class interval is 8. So, an 8 goes in the column marked Frequency. There's your frequency distribution.

> Sometimes, it is a good idea to graph your data first and then do whatever calculations or analysis is called for. By first looking at the data, you may gain insight into the relationship between variables, what kind of descriptive statistic is the right one to use to describe the data, and so on. This extra step might increase your insights and the value of what you are doing.

THE PLOT THICKENS: CREATING A HISTOGRAM

Now that we've got a tally of how many scores fall in which class intervals, we'll go to the next step and create what is called a **histogram,** which is a visual representation of the frequency distribution where the frequencies are represented by bars.

> Depending on the book you read and the software you use, visual representations of data are called graphs or charts. It really makes no difference. All you need to know is that a graph or a chart is the visual representation of data.

To create a histogram by hand, do the following.

1. Using a piece of graph paper, place values at equal distances along the x-axis, as shown in Figure 4.1. Now, identify the midpoint of the class intervals, which is the middle point in the class interval.

It's pretty easy to just eyeball, but you can also just add the top and bottom values of the class interval and divide by 2. For example, the midpoint of the class interval 0–4 is the average of 0 and 4, or 4/2 = 2.

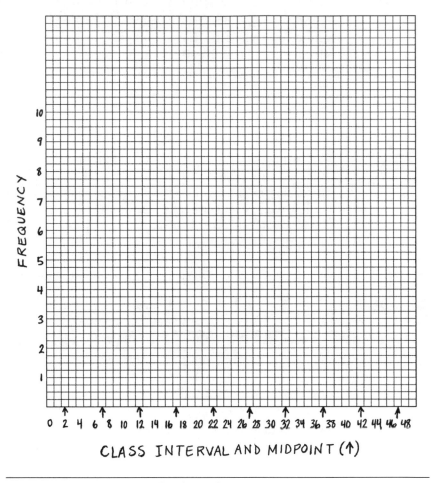

Figure 4.1   Class Intervals Along the *x*-Axis

2. Draw a bar or column around each midpoint that represents the entire class interval to the height representing the frequency of that class interval. For example, in Figure 4.2, you can see our first entry where the class interval of 0–4 is represented by the frequency of 1 (representing the one time a value between 0 and 4 occurs). Continue drawing bars or columns until each of the frequencies for each of the class intervals is represented.

Here's a nice hand-drawn (really!) histogram for the frequency distribution of 50 scores with which we have been working so far.

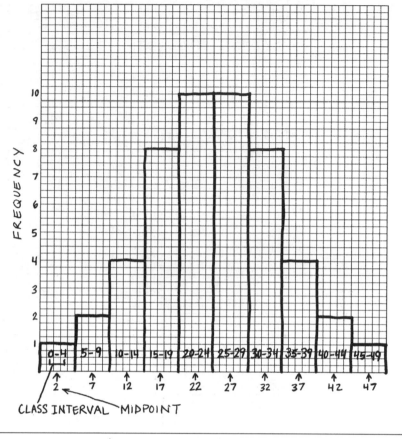

Figure 4.2 A Hand-Drawn Histogram

Notice how each class interval is represented by a range of scores along the *x*-axis and is separately the size of a class interval, which in this case is 5.

The Tally-Ho Method

You can see by the simple frequency distribution that you saw at the beginning of the chapter that you already know more about the distribution of scores than just a simple listing of them.

You have a good idea of what values occur with what frequency. But another visual representation (besides a histogram) can be done by using tallies for each of the occurrences, as shown in Figure 4.3.

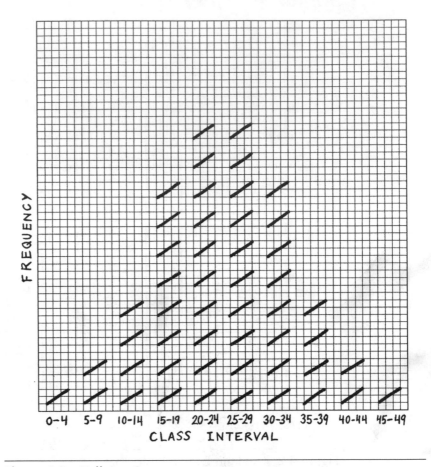

Figure 4.3   Tallying Scores

We used tallies that correspond with the frequency of scores that occur within a certain class. This gives you an even better visual representation of how often certain scores occur relative to other scores.

Using the Amazing Analysis ToolPak to Create a Histogram

Here's how Excel creates a histogram using the Histogram tool in the Analysis ToolPak.

In sum, you will need to do three things:

1. Identify the cells that contain the data from which the histogram will be created.

2. Identify the "bins" or containers (we called them class intervals earlier in this chapter) in which you want to put the data.

3. Select the Histogram tool from the ToolPak and you're in business.

To create a histogram using the ToolPak, follow these steps. Enter the data in a new worksheet (as you see in Figure 4.4).

	A	B	C	D	E
1	Scores				
2	47	10	31	25	20
3	2	11	31	25	21
4	44	14	15	26	21
5	41	14	16	26	21
6	7	30	17	27	24
7	6	30	16	29	24
8	35	32	15	29	23
9	38	33	19	28	20
10	35	34	18	29	21
11	36	32	16	27	20

Figure 4.4   The Data Being Used to Create a Histogram

1. Enter the bins you want to use (as you can in Figure 4.5). Instead of the range of a bin (or a class interval), you just enter the starting point (such as 4), and the next bin becomes the starting point for the next class interval (such as 9). You can see these bins in Cells G2 through G11 with a title that we entered as well.

	A	B	C	D	E	F	G
1	Scores						Bins
2	47	10	31	25	20		49
3	2	11	31	25	21		44
4	44	14	15	26	21		39
5	41	14	16	26	21		34
6	7	30	17	27	24		29
7	6	30	16	29	24		24
8	35	32	15	29	23		19
9	38	33	19	28	20		14
10	35	34	18	29	21		9
11	36	32	16	27	20		4

Figure 4.5   Creating Bins for the Histogram ToolPak

2. Highlight any blank cell in the worksheet. That's where the histogram will be placed.

3. Click Data → Data Analysis, and you will see the Data Analysis dialog box as shown in Figure 4.6.

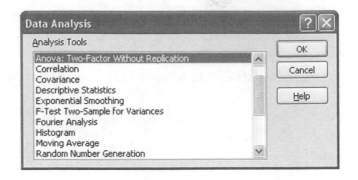

Figure 4.6   The Data Analysis Dialog Box

4. Scroll down until you see the Histogram tool, and double-click on it. When you do this, you will see the Histogram dialog box as shown in Figure 4.7.

Figure 4.7   The Histogram Dialog Box

5. Click on the Input Range box and drag the mouse over the range that contains the data, which in this case is A2 through E11.

6. Click on the Bin Range and drag the mouse over the range that contains the bins, which in this case is G2 through G11.

7. Click the Output Range button. Enter the address where the histogram will be created. One cell will do. In our example, we entered H1. The completed dialog box for creating a histogram is shown in Figure 4.8.

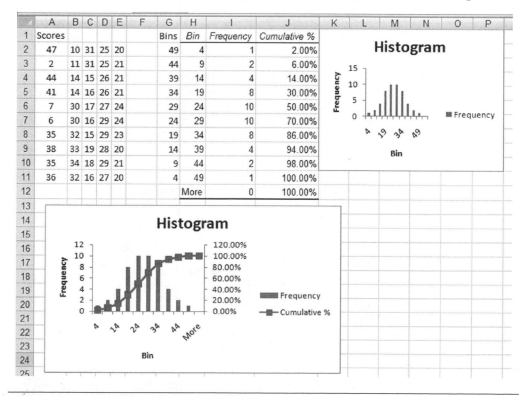

Figure 4.8   The Completed Histogram Dialog Box

8. Click the Cumulative Percentage and Chart Output options. The first gives you the cumulative values (adding up the frequencies) and the second a visual view.

9. Click OK, and there you go as you see in Figure 4.9: a pretty histogram.

Notice a few things about the histogram you see in Figure 4.9.

	A	B	C	D	E	F	G	H	I	J
1	Scores						Bins	*Bin*	*Frequency*	*Cumulative %*
2	47	10	31	25	20		49	4	1	2.00%
3	2	11	31	25	21		44	9	2	6.00%
4	44	14	15	26	21		39	14	4	14.00%
5	41	14	16	26	21		34	19	8	30.00%
6	7	30	17	27	24		29	24	10	50.00%
7	6	30	16	29	24		24	29	10	70.00%
8	35	32	15	29	23		19	34	8	86.00%
9	38	33	19	28	20		14	39	4	94.00%
10	35	34	18	29	21		9	44	2	98.00%
11	36	32	16	27	20		4	49	1	100.00%
12								More	0	100.00%

Figure 4.9   The Finished ToolPak Histogram

First, Excel always creates bins with the lowest bin value first.

Second, it creates two columns, one named Bin (where it creates the class intervals), and one named Frequency (where it enters the frequencies).

Third, it creates a category labeled More, where any values outside of the range that the bins can contain are placed. So, for example, if there was a value such as 87 in the data set, More would list that as 1.

The Next Step: A Frequency Polygon

Creating a histogram or a tally of scores wasn't so difficult, and the next step (and the next way of illustrating data) is even easier. We're going to use the same data—and, in fact, the histogram that you just saw created in Figure 4.2—to create a frequency polygon. A **frequency polygon** is a continuous line that represents the frequencies of scores within a class interval, as shown in Figure 4.10.

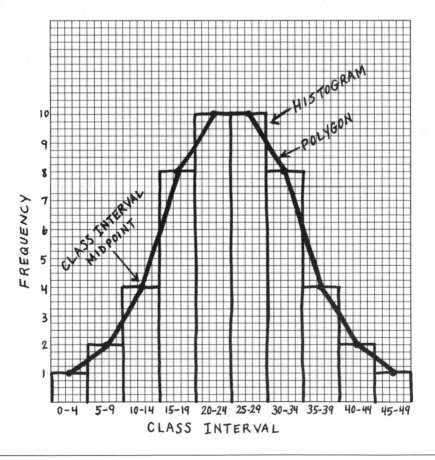

Figure 4.10   A Hand-Drawn Frequency Polygon

How did we draw this? Here's how.

1. Place a midpoint at the top of each bar or column in a histogram (see Figure 4.2).
2. Connect the lines, and you've got it—a frequency polygon!

Why use a frequency polygon rather than a histogram to represent data? It's more a matter of preference than anything else. A frequency polygon appears more dynamic than a histogram (a line that represents change in frequency always looks neat), but you are basically conveying the same information.

Cumulating Frequencies

Once you have created a frequency distribution and have visually represented those data using a histogram or a frequency polygon, another option is to create a visual representation of the cumulative frequency of occurrences by class intervals. This is called a **cumulative frequency distribution.**

A cumulative frequency distribution is based on the same data as a frequency distribution, but with an added column (Cumulative Frequency), as shown below.

Class Interval	Frequency	Cumulative Frequency
45–49	1	50
40–44	2	49
35–39	4	47
30–34	8	43
25–29	10	35
20–24	10	25
15–19	8	15
10–14	4	7
5–9	2	3
0–4	1	1

The cumulative frequency distribution begins by the creation of a new column labeled Cumulative Frequency. Then, we add the frequency in a class interval to all the frequencies below it.

For example, for the class interval of 0–4, there is 1 occurrence and none below it, so the cumulative frequency is 1. For the class interval of 5–9, there are 2 occurrences in that class interval and 1 below it for a total of 3 (2 + 1) occurrences in that class interval or below it. The last class interval (45–49) contains 1 occurrence, and there is a total of 50 occurrences at or below that class interval.

Once we create the cumulative frequency distribution, then the data can be plotted just as they were for a histogram or a frequency polygon. Only this time, we'll skip right ahead and plot the midpoint of each class interval as a function of the cumulative frequency of that class interval. You can see the cumulative frequency distribution in Figure 4.11 based on the 50 scores from the beginning of this chapter.

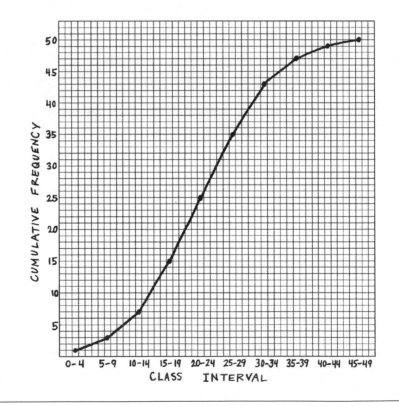

Figure 4.11   A Cumulative Frequency Distribution Drawn by Hand

TECH TALK   Another name for a cumulative frequency polygon is an **ogive**. And, if the distribution of the data is normal (see Chapter 7 for more on this), then the ogive represents another way to illustrate what is popularly known as a bell curve or a normal distribution.

FAT AND SKINNY FREQUENCY DISTRIBUTIONS

You could certainly surmise by now that distributions can be very different from one another in a variety of ways. In fact, there are four different ways: average value, variability, skewness, and kurtosis. Those last two are new terms, and we'll define them as we show you what they look like. Let's define each of the four characteristics and then illustrate them.

Average Value

We're back once again to measures of central tendency. You can see in Figure 4.12 how three different distributions can differ in their average value. Notice that the average for Distribution C is more than the average for Distribution B, which, in turn, is more than the average for Distribution A.

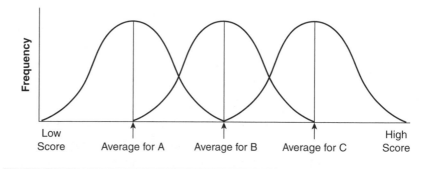

Figure 4.12   How Distributions Can Differ in Their Average Score

Variability

In Figure 4.13, you can see three distributions that all have the same average value but differ in variability. The variability in Distribution A is less than that in Distribution B and, in turn, less than that found in C. Another way to say this is that Distribution C has the largest amount of variability of the three distributions, and A has the least.

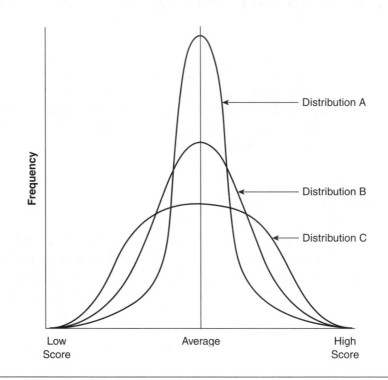

Figure 4.13   How Distributions Can Differ in Variability

Skewness

Skewness is a measure of the lack of symmetry, or the lopsidedness, of a distribution. In other words, one "tail" of the distribution is longer than another. For example, in Figure 4.14, Distribution A's right tail is longer than its left tail, which corresponds to a smaller number of occurrences at the high end of the distribution. This is a positively skewed distribution. This might be the case when you have a test that is very difficult, and few people get scores that are relatively high and many more get scores that are relatively low.

Distribution C's right tail is shorter than its left tail, which corresponds to a larger number of occurrences at the high end of the distribution. This is a negatively skewed distribution and would be the case for an easy test (lots of high scores and relatively few low scores). And Distribution B—well, it's just right, equal lengths of tails and no skewness. If the mean is greater than the median, the distribution is positively skewed. If the median is greater than the mean, the distribution is negatively skewed.

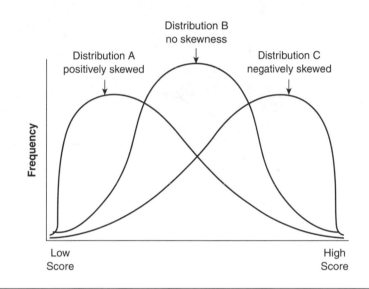

Figure 4.14   Degree of Skewness in Different Distributions

Kurtosis

Even though this sounds like a medical condition, it's the last of the four ways that we can classify how distributions differ from one another. **Kurtosis** has to do with how flat or peaked a distribution appears, and the terms used to describe this characteristic are relative ones. For example, the term **platykurtic** refers to a distribution that is relatively flat compared to a normal, or bell-shaped, distribution. The term **leptokurtic** refers to a distribution that is relatively peaked compared to a normal, or bell-shaped, distribution. In Figure 4.14, Distribution A is platykurtic compared to Distribution B. Distribution C is leptokurtic compared to Distribution B. Figure 4.15 looks similar to Figure 4.13 for a good reason—distributions that are platykurtic, for example, are relatively more dispersed than those that are not. Similarly, a distribution that is leptokurtic is less variable or dispersed relative to others.

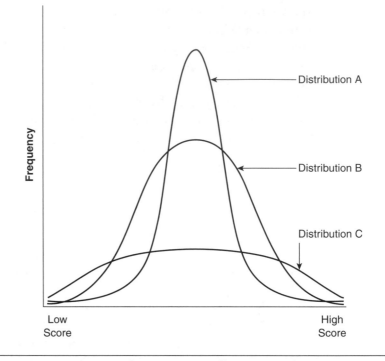

Figure 4.15   Degrees of Kurtosis in Different Distributions

More Excel

Excel can also help you compute a value for kurtosis and skewness of a set of scores. You can use the **SKEW** and the **KURT** functions and compute such values. For example, if you have 20 scores in Cells A1 through A20, then the functions would look like this:
=SKEW(A1:A20) and
=KURT(A1:A20).
Just enter the function in the cell where you want the results returned.

TECH TALK   While skewness and kurtosis are used mostly as descriptive terms (such as "That distribution is negatively skewed"), there are mathematical indicators of how skewed or kurtotic a distribution is. For example, skewness is computed by subtracting the value of the median from the mean. For example, if the mean of a distribution is 100 and the median is 95, the skewness value is $100 - 95 = 5$, and the distribution is positively skewed.

If the mean of a distribution is 85 and the median is 90, the skewness value is $85 - 90 = -5$, and the distribution is negatively skewed. There's an even more sophisticated formula, which is not relative but takes the standard deviation of the distribution into account so that skewness indicators can be compared to one another (see Formula 4.1):

$$Sk = \frac{3(\bar{X} - M)}{s},$$ (4.1)

Sk = is Pearson's (he's the correlation guy you'll learn about in Chapter 5) measure of skewness,

$\bar{X}$ = is the mean, and

M = is the median.

Using this formula, we can compare the skewness of one distribution to another in absolute, and not relative, terms. For example, the mean of Distribution A is 100, the median is 105, and the standard deviation is 10. For Distribution B, the mean is 120, the median is 116, and the standard deviation is 10. Using Pearson's formula, the skewness of Distribution A is –1.5, and the skewness of Distribution B is 1.2. Distribution A is negatively skewed, and Distribution B is positively skewed. However, Distribution A is more skewed than Distribution B, regardless of the direction.

EXCELLENT CHARTS

Now the fun begins.

In this section of this chapter on charts, we'll show you how Excel can be used to create a simple chart in a matter of minutes (actually seconds, and only three, that's right, three steps) and, once created, how you can modify it to pretty it up and make it even more informative. We'll create the column chart you see in Figure 4.16 that shows the number of males and females who participated in a survey, then show you how you can make some simple, but impressive, edits.

OK—so the chart in Figure 4.16 looks like two bars and not two columns. The folks who created Excel named charts with bars that are vertical as column charts and charts with bars that are horizontal as bar charts. That's just the way it is.

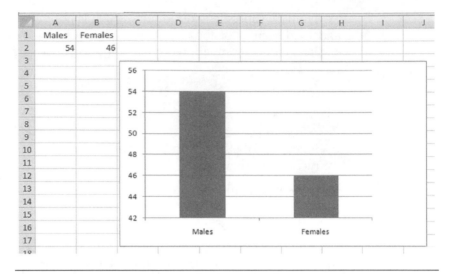

Figure 4.16   Creating a Simple Column Chart

Just a few things to remember about creating Excel charts:

Excel will chart the data that you highlight, but you may have to first compute the values you want charted. For example, if you have a list of values and want to chart the average value, you will have to first compute the average value and then chart that.

Excel creates a chart of the worksheet that contains the data used to create that chart in the first place, as you see in Figure 4.16. When you save a worksheet that contains a chart, you save both the data on which the chart is based as well as the chart itself.

When the data that were used to create a chart change, the chart itself will change. These two elements (the data and the chart) are linked.

Charts can easily be cut and pasted into other applications (as we will show you later).

Excel offers a myriad (that means lots and lots) of different charts, all of which can be created with just a few clicks and then modified as you see fit.

Your First Excel Chart: A Moment to Remember

To create a chart in Excel, follow these steps:

1. Highlight the data you want to use to create the chart. In Figure 4.16, we would highlight Cells A1 through B2.

2. Click the Insert tab, and in the Charts group (which you can see in Figure 4.17), select the type of chart you want to use. In Figure 4.17, we selected Column.

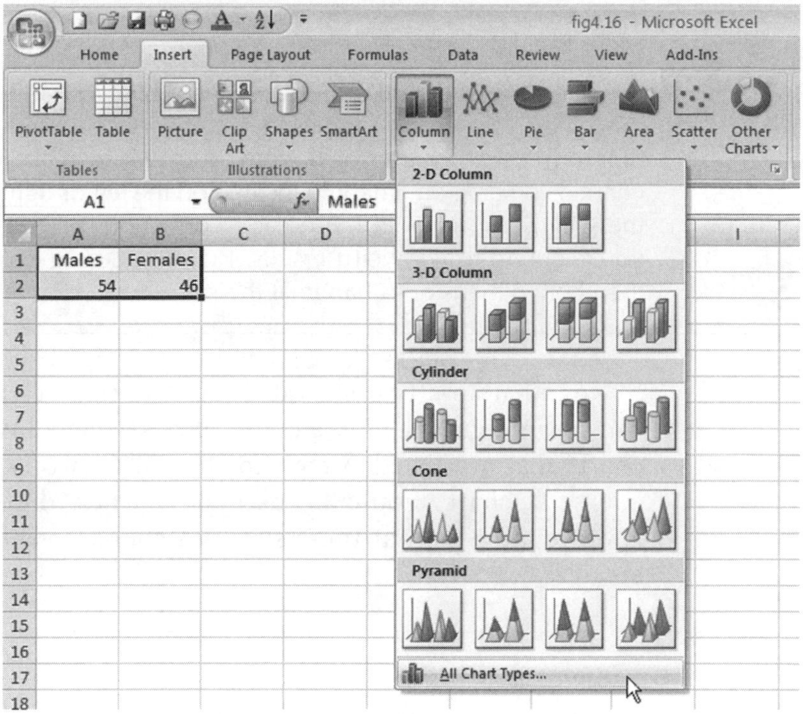

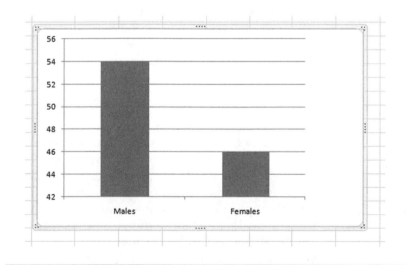

Figure 4.17 Selecting a Specific Type of Chart to Create

3. Click the first icon under 2-D Column (because Column will already be selected), and zingo—there you have it as shown in Figure 4.18. A very nice and tidy column chart completed in very few clicks. It just does not get any easier.

Figure 4.18 A Simple Column Chart Created Using Excel

More Excel

You can easily increase or decrease the size of a chart on a worksheet by dragging on any of the handles that appear once a chart is selected. If you drag on one at the top or bottom, you will increase the height of the chart. If you drag on a handle located on the left or right, you will change the width. But, in both these cases, only one dimension changes, and you're bound to get weird results. But, if you drag on a corner handle, then both dimensions change at the same time. Wonderful.

Now, that's pretty easy and all done in three clicks. And, as you can see in Figure 4.17, we could have selected any of the types of charts that are listed (or go to the All Chart Types option). Of course, a chart should be used only when the data are of the appropriate type, but there sure is a cool assortment of charts to create.

More Excel

But what if you want to place a chart on its own worksheet? Here's how:

1. Click on the body of the chart to open the Chart Tools set of tabs.

2. Click Move Chart (on the right-hand side of the Chart Tools set of tabs).

3. Now you can select to move the chart to a new worksheet (and you have to provide a name for this new sheet), or you can move it to another, already created worksheet. Remember, that if you create a new one, then a new tab will appear in the workbook (and that new tab is for the new sheet you created). That's where the chart will be.

TECH TALK

Once you create a chart, it is like any other Excel element and can be easily cut or copied from Excel and pasted into a new document (such as Word or any other Windows or Mac application). However, if you just do the straight cut or copy and paste, then changes in the data used to create the chart will not appear on the chart that is pasted into another application. To do this, you have to dynamically link the data used to create the chart to the chart where it will appear. This is very cool stuff, but be careful because you must remember that changes in your Excel worksheet will result in changes in the chart no matter where it is pasted. Follow these steps:

1. Highlight the chart. Right click and select the Copy option.

2. Switch to the Word document into which you want to paste the chart (but it can be any other Microsoft Office product such as PowerPoint) and often non-Microsoft products as well.

3. On the Home tab, click Paste, and the chart will appear in the Word document.

4. In the lower right-hand corner of the chart will appear a small Paste Options icon as you see here: 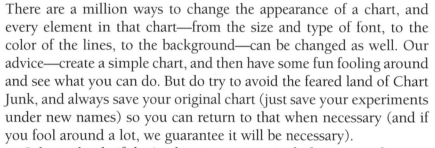. Click on that and select Keep Source Formatting. This way, whatever changes are made in the chart created in Excel will also be made in the chart appearing in the Word document.

EXCELLENT CHARTS PART DEUX: MAKING CHARTS PRETTY

There are a million ways to change the appearance of a chart, and every element in that chart—from the size and type of font, to the color of the lines, to the background—can be changed as well. Our advice—create a simple chart, and then have some fun fooling around and see what you can do. But do try to avoid the feared land of Chart Junk, and always save your original chart (just save your experiments under new names) so you can return to that when necessary (and if you fool around a lot, we guarantee it will be necessary).

Color is dandy if that's what you want to include in your charts to further emphasize a particular visual point. However, color can be a mess unless you use it sparingly; also, color sometimes prints pretty weirdly on black-and-white printers. So, stay simple and use color only when you need to and when you can print in color as well.

To modify the appearance of the chart you saw in Figure 4.18, follow these steps. The general rule for changing the appearance of an element in a chart is to double-click on the element and then make any adjustments you want. Remember that clicking once on the chart itself also makes the Chart Tools tabs available for changes. When the Chart Tools tabs are exposed, you will see tabs for Design, Layout, and Format.

1. Click on the chart to highlight it so that all the Chart Tools are available.

2. Select the Layout tab, then the Chart Title option, and then the Above Chart option to place the title above the chart as you see in Figure 4.19 (where we already entered the title).

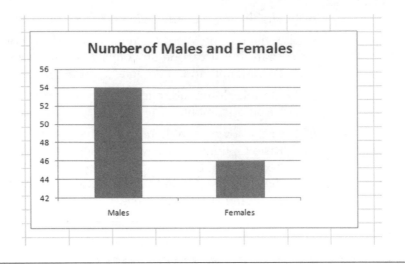

Figure 4.19   Entering a Title for a Chart

3. Then, get rid of the horizontal gridlines in the chart by selecting the Layout tab, then the Gridlines option, then Primary Horizontal Gridlines, and then None. You can also just click on any one and press the Delete key on your keyboard!

4. Next, click on the Series label and press the Delete key because we don't need that on the chart.

5. Finally, we can change the pattern that is used to illustrate the bars. To do this, right click on the bar you want to modify, select Format Data Series, select Fill from the Format Data Series dialog box, click on Solid Fill, and select the color you want from the drop-down menu as you see in Figure 4.20.

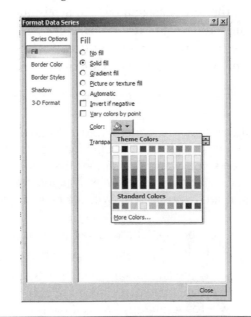

Figure 4.20   Changing the Fill in Each Chart Element

As you can see, you can pick a variety of fills, colors, styles, shadows, and more for the chart that you are modifying.

Our advice—fool around with all these options because it is the best way to get to know what they can do and when you should use them.

OTHER COOL CHARTS

What we did so far in this chapter is take some data and show how charts such as histograms and polygons can be used to communicate visually. But there are several other types of charts that are used in the behavioral and social sciences, and although it's not necessary for you to know exactly how to create them, you should at least be familiar with their names and what they do. So, here are some popular charts and when to use them.

Bar Charts

A bar chart is identical to a column chart, but in this chart, categories are organized on the *y*-axis and values are shown horizontally on the *x*-axis.

Line Charts

A line chart should be used when you want to show a trend in the data at equal intervals. Here are some examples of when you might want to use a line chart:

- Number of cases of mononucleosis (mono) per season among college students at three state universities
- Change in student enrollment over the school year
- Number of travelers on two different airlines for each quarter

In Figure 4.21, you can see a chart of the number of reported cases of mono by season among college students at three state universities.

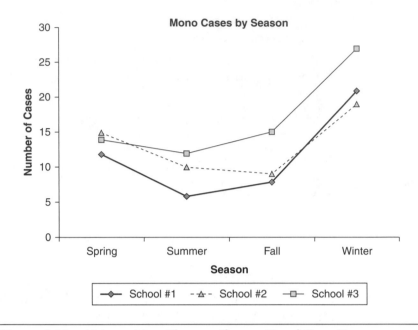

Figure 4.21   Using a Line Chart to Show a Trend Over Time

Pie Charts

A pie chart should be used when you want to show the proportion of an item that makes up a series of data points. Here are some examples of when you might want to use a pie chart:

- Percentage of children living in poverty by ethnicity
- Proportion of night students and day students enrolled
- Age of participants by gender

In Figure 4.22, you can see a pie chart of the number of voters by political party. We made it three-dimensional (sort of just for fun), but it doesn't look so bad in any case.

PIVOT THIS!

Time for something really new—pivot tables.

A **pivot table** allows you to summarize information that you find in lists and tables without using any formulas or functions. It has the word *pivot* in the title because you rearrange or rotate information such as that contained in rows and columns (as many Excel worksheets are organized).

Now, the creation of these can be quite intimidating, and that's one reason why people stay away from them. But, if we start simple, we can

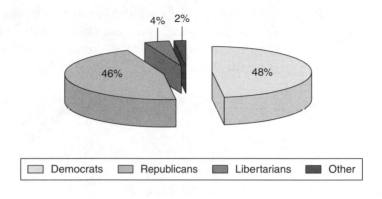

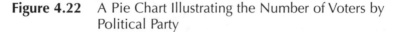

Figure 4.22 A Pie Chart Illustrating the Number of Voters by Political Party

create a useful one and demonstrate some of the basics—from here, you can expand your set of skills by creating and using more complex tables.

Here's a very simple example. We have a list of 20 individuals and two data points on each individual, as you see in Figure 4.23 (and we have already selected the data we want to use to create the pivot table). The first variable is gender (1 = male and 2 = female), and the second is category of program in which they are studying (A, B, or C).

To create the table, follow these steps:

1. Select the data that you want to include in the table. In this example, the data are located in Cells B1 through C21. We don't want to use the ID column.

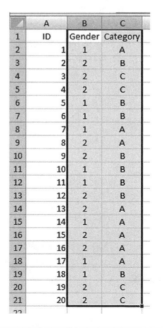

	A	B	C
1	ID	Gender	Category
2	1	1	A
3	2	2	B
4	3	2	C
5	4	2	C
6	5	1	B
7	6	1	B
8	7	1	A
9	8	2	A
10	9	2	B
11	10	1	B
12	11	1	B
13	12	2	B
14	13	2	A
15	14	1	A
16	15	2	A
17	16	2	A
18	17	1	A
19	18	1	B
20	19	2	C
21	20	2	C

Figure 4.23 Data for a Pivot Table

2. On the Insert tab, click the PivotTable option on the left-hand side of the screen, and then click the PivotTable item on the drop-down menu. When you do that, you will see the Create PivotTable dialog box as you see in Figure 4.24.

Figure 4.24   The Create PivotTable Dialog Box

3. Select the location where you want the new table to appear. In this example, we placed it on the same worksheet where the data are located.

4. Click OK, and you will see the beginnings of the table and the Pivot Table Field List (on the right) as shown in Figure 4.25.

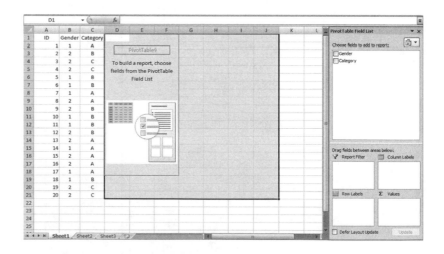

Figure 4.25   The Pivot Table Field List

This is the list into which you drag column labels to rearrange them.

5. Drag the Gender field to the Row Labels box in the Pivot Table Field List. When you do that, you will see the table being formed in the main area of the worksheet, as shown in Figure 4.26.

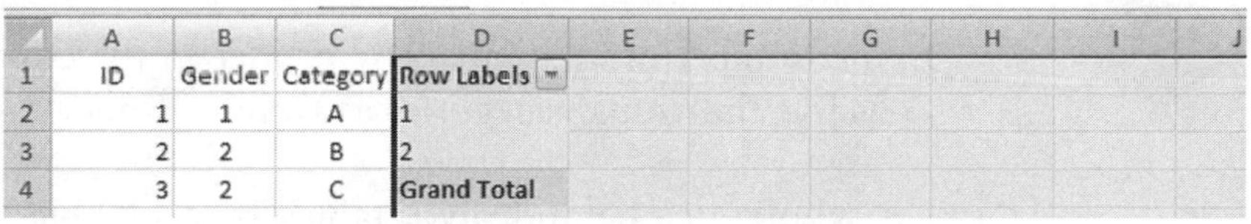

	A	B	C	D	E	F	G	H	I	J
1	ID	Gender	Category	Row Labels ▾						
2	1	1	A	1						
3	2	2	B	2						
4	3	2	C	Grand Total						

Figure 4.26   Creating the Pivot Table by Dragging Fields

6. Now, drag the Category field to the column category.

7. Finally, drag the Category fields to the Σ Values area in the lower right part of the Pivot Table Field List, and as a result, you will see the finished Pivot Table in Figure 4.27 (a bit cleaned up so it would all fit on one screen).

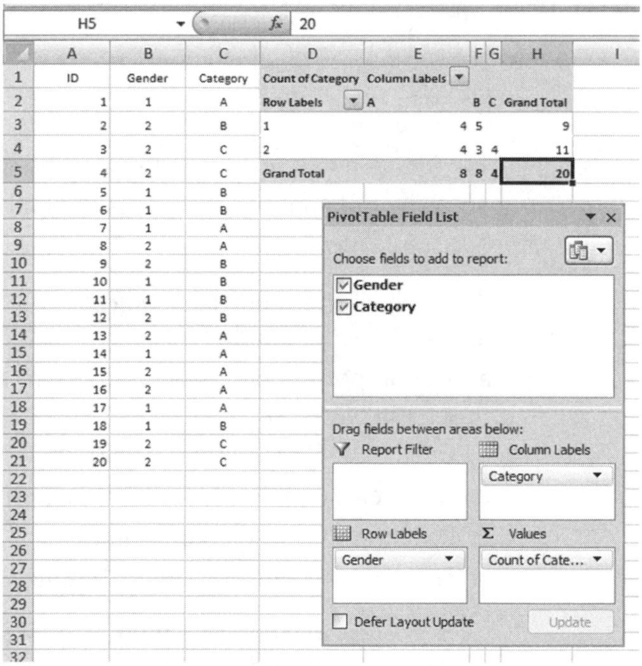

Figure 4.27   A Completed Simple Pivot Table

OK—not so bad. What do we have that we didn't have before? With very few clicks, we can now see:

1. the number of individuals in each Gender (nine in Category 1 and 11 in Category 2),

2. the number of individuals in each Category (eight in A, eight in B, and four in C), and

3. the number of individuals by gender in the entire group (such as five individual categories as Gender 1 and Category B).

The whole idea behind pivot tables is to be able to take already created data and create tables that extract exactly the information you want to use.

Summary

There's no question that charts are fun to create and can add enormous understanding to what appear to be disorganized data. Follow our suggestions in this chapter and use charts well, but only when they enhance, not just add to, what's already there.

Time to Practice

1. A data set of 50 comprehension scores (Comp Score) named Chapter 4 Data Set 1 is available on the Web site and in Appendix C. Answer the following questions and/or complete the following tasks:

   1a. Create a frequency distribution and a histogram for the set.

   1b. Why did you select the class interval you used?

   1c. Is this distribution skewed? How do you know?

2. For each of the following, indicate whether you would use a pie, line, or bar chart, and why:

   a. The proportion of freshmen, sophomores, juniors, and seniors in a particular university

   b. Change in GPA over four semesters

   c. Number of applicants for four different jobs

   d. Reaction time to different stimuli

   e. Number of scores in each of 10 categories

3. Go to the library and find a journal article in your area of interest that contains empirical data, but does not contain any visual representation of them. Use the data to create a chart. Be sure to specify what type of chart you are creating, and why you chose the one you did. You can create the chart manually or use Excel.

4. Create the worst-looking chart that you can, crowded with chart and font junk. There's nothing that has as lasting an impression as a *bad* example.

5. Identify at least one element that could be modified in the pie chart you see in Figure 4.22.

Answers to Practice Questions

1a. Here's the frequency distribution.

Class Interval	Frequency
45–49	1
40–44	2
35–39	3
30–34	8
25–29	10
20–24	10
15–19	8
10–14	4
5–9	2
0–4	2

Here's what the histogram (done using Excel) should look like.

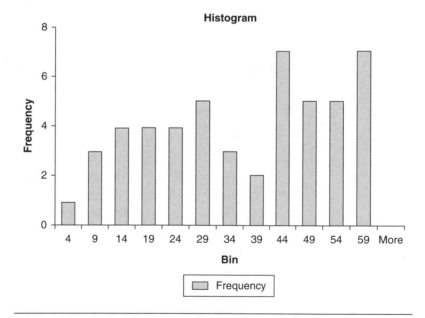

Figure 4.28 Histogram of Data in Chapter 4 Data Set 1

1b. We settled on a class interval of 5 because it allowed us to have close to 10 class intervals, and it fit the criteria that we discussed in this chapter for deciding on a class interval.

1c. The distribution is negatively skewed because the mean is less than the median.

2a. Pie

2b. Line

2c. Bar or Column

2d. Line

2e. Bar or Column

3. On your own!

4. We did this using Excel and the chart editor, and it's as uninformative as it is ugly. It's the pie chart we showed you earlier (Figure 4.22) revised by Dr. Frankenstein and now appearing as Figure 4.29.

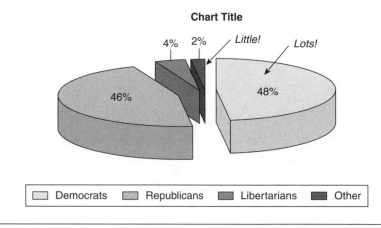

Figure 4.29   A Really, Really Ugly Chart Brought to You by Chart Junk

5. Most obvious is that the 3D aspect of the chart can be dumped—it does not add anything to the information conveyed. Also, the chart could use a title and a subtitle, and the segments could be more easily distinguished from one another.

5 Ice Cream and Crime

Computing Correlation Coefficients

Difficulty Scale ☺☺ (moderately hard)

How much Excel? ▦ ▦ ▦ ▦ (lots and lots)

What you'll learn about in this chapter

- What correlations are and how they work
- How to compute a simple correlation coefficient
- Using the CORREL function to compute a correlation
- Using the Correlation tool in the Analysis ToolPak to compute a correlation and a correlation matrix
- How to interpret the value of the correlation coefficient
- What other types of correlations exist and when to use them

WHAT ARE CORRELATIONS ALL ABOUT?

Measures of central tendency and measures of variability are not the only descriptive statistics that we are interested in using to get a picture of what a set of scores looks like. You have already learned that knowing the values of the one most representative score (central tendency) and that a measure of spread or dispersion (variability) is critical for describing the characteristics of a distribution.

However, sometimes we are as interested in the relationship between variables—or, to be more precise, how the value of one variable changes when the value of another variable changes. The way we express this interest is through the computation of a simple correlation coefficient.

113

A **correlation coefficient** is a numerical index that reflects the relationship between two variables. The value of this descriptive statistic ranges between a value of −1 and a value of +1. A correlation between two variables is sometimes referred to as a **bivariate** (for two variables) **correlation.** Even more specifically, the type of correlation that we talk about in the majority of this chapter is called the **Pearson product-moment correlation,** named for its inventor, Karl Pearson.

TECH TALK The Pearson correlation coefficient examines the relationship between two variables, when both of those variables are continuous in nature. In other words, they are variables that can assume any value along some underlying continuum, such as height, age, test score, or income. But there is a host of other variables that are not continuous. They're called discrete or categorical variables, like race (such as black and white), social class (such as high and low), and political affiliation (such as Democrat and Republican). You need to use other correlational techniques, such as the point-biserial correlation, in these cases. These topics are for a more advanced course, but you should know that they are acceptable and very useful techniques. We mention them briefly later in this chapter.

There are other types of correlation coefficients that measure the relationship between more than two variables, and we'll leave those for the next statistics course (which you are looking forward to already, right?).

Types of Correlation Coefficients: Flavor 1 and Flavor 2

A correlation reflects the dynamic quality of the relationship between variables. In doing so, it allows us to understand whether variables tend to move in the same or opposite directions when they change. If variables change in the same direction, the correlation is called a **direct correlation** or a positive correlation. If variables change in opposite directions, the correlation is called an **indirect correlation** or a negative correlation. Table 5.1 shows a summary of these relationships.

Now, keep in mind that the examples in the table reflect generalities. For example, regarding time to completion and the number of items correct on a test, in general, the less time that is taken on a test, the lower the score.

Table 5.1   Types of Correlations and the Corresponding Relationship
Between Variables

What Happens to Variable X	What Happens to Variable Y	Type of Correlation	Value	Example
X increases in value	Y increases in value	Direct or positive	Positive, ranging from .00 to +1.00	The more time you spend studying, the higher your test score will be.
X decreases in value	Y decreases in value	Direct or positive	Positive, ranging from .00 to +1.00	The less money you put in the bank, the less interest you will earn.
X increases in value	Y decreases in value	Indirect or negative	Negative, ranging from −1.00 to .00	The more you exercise, the less you will weigh.
X decreases in value	Y increases in value	Indirect or negative	Negative, ranging from −1.00 to .00	The less time you take to complete a test, the more you'll get wrong.

Such a conclusion is not rocket science, because the faster one goes, the more likely one is to make careless mistakes, such as not reading instructions correctly. But, of course, there are people who can go very fast and do very well. And there are people who go very slow and don't do well at all. The point is that we are talking about the performance of a group of people on two different variables. We are computing the correlation between the two variables for the group, not for any one particular person.

THINGS TO REMEMBER

There are several (easy but important) things to remember about the correlation coefficient:

A correlation can range in value from −1 to +1.

The absolute value of the coefficient reflects the strength of the correlation. So, a correlation of −.70 is stronger than a correlation of +.50. One of the frequently made mistakes regarding correlation coefficients occurs when students assume that a direct or positive correlation is always stronger (i.e., "better") than an indirect or negative correlation because of the sign and nothing else. Ain't so. A correlation always reflects the situation where there are at least two data points (or variables) per case.

Another easy mistake is to assign a value judgment to the sign of the correlation. Many students assume that a negative relationship is not good and a positive one is good. That's why, instead of using the terms "negative" and "positive," the terms "indirect" and "direct" communicate meaning more clearly.

The Pearson product-moment correlation coefficient is represented by the small letter r with a subscript representing the variables that are being correlated. For example,

r_{xy} is the correlation between variable X and variable Y,

$r_{weight \cdot height}$ is the correlation between weight and height,

$r_{SAT \cdot GPA}$ is the correlation between SAT score and grade point average (GPA).

 TECH TALK The correlation coefficient reflects the amount of variability that is shared between two variables and what they have in common. For example, you can expect an individual's height to be correlated with an individual's weight because the two variables share many of the same characteristics, such as the individual's nutritional and medical history, general health, and genetics. However, if one variable does not change in value and therefore has nothing to share, then the correlation between the two variables is zero. For example, if you computed the correlation between age and number of years of school completed, and everyone was 25 years old, there would be no correlation between the two variables because there's literally nothing (any variability) about age available to share.

Likewise, if you constrain or restrict the range of one variable, the correlation between that variable and another variable is going to be less than if the range is not constrained. For example, if you correlate reading comprehension and grades in school for very high-achieving children, you'll find the correlation lower than if you computed the same correlation for children in general. That's because the reading comprehension score of very high-achieving students is quite high and much less variable than it would be for all children. The moral? When you are interested in the relationship between two variables, try to collect sufficiently diverse data—that way, you'll get the truest and most representative result.

COMPUTING A SIMPLE CORRELATION COEFFICIENT

The computational formula for the simple Pearson product moment correlation coefficient between a variable labeled X and a variable labeled Y is shown in Formula 5.1.

$$r_{xy} = \frac{n\Sigma XY - \Sigma X \Sigma Y}{\sqrt{[n\Sigma X^2 - (\Sigma X)^2][n\Sigma Y^2 - (\Sigma Y)^2]}},$$ (5.1)

where

r_{xy} is the correlation coefficient between X and Y;

n is the size of the sample;

X is the individual's score on the X variable;

Y is the individual's score on the Y variable;

XY is the product of each X score times its corresponding Y score;

X^2 is the individual X score, squared; and

Y^2 is the individual Y score, squared.

Here are the data we will use in this example:

X	Y	X^2	Y^2	XY
2	3	4	9	6
4	2	16	4	8
5	6	25	36	30
6	5	36	25	30
4	3	16	9	12
7	6	49	36	42
8	5	64	25	40
5	4	25	16	20
6	4	36	16	24
7	5	49	25	35
Total, Sum, or Σ 54	43	320	201	247

Before we plug the numbers in, let's make sure you understand what each one represents.

ΣX, or the sum of all the X values, is 54;

ΣY, or the sum of all the Y values, is 43;

ΣX^2, or the sum of each X value squared, is 320;

ΣY^2, or the sum of each Y value squared, is 201; and

ΣXY, or the sum of the products of X and Y, is 247.

It's easy to confuse the sum of a set of values squared and the sum of the squared values. The sum of a set of values squared is taking values such as 2 and 3, summing them (to be 5), and then squaring that (which is 25). The sum of the squared values is taking values such as 2 and 3, squaring them (to get 4 and 9, respectively), and then adding those together (to get 13). Just look for the parentheses as you work.

Here are the steps in computing the correlation coefficient:

1. List the two values for each participant. You should do this in a column format so as not to get confused.

2. Compute the sum of all the X values, and compute the sum of all the Y values.

3. Square each of the X values, and square each of the Y values.

4. Find the sum of the XY products.

These values are plugged into the equation you see in Formula 5.2:

$$r_{xy} = \frac{(10 \times 247) - (54 - 43)}{\sqrt{[(10 \times 320) - 54^2][(10 \times 201) - 43^2]}}. \tag{5.2}$$

Ta-da! And you can see the answer in Formula 5.3:

$$r_{xy} = \frac{148}{213.83} = .692. \tag{5.3}$$

And Now . . . Using Excel's CORREL Function

To compute the correlation between two variable numbers using Excel, follow these steps.

1. Enter the individual scores into one column in a worksheet, such as you see in Figure 5.1.

fig5.1	A	B
	X	Y
1	X	Y
2	2	3
3	4	2
4	5	6
5	6	5
6	4	3
7	7	6
8	8	5
9	5	4
10	6	4
11	7	5

Figure 5.1 Data for Computing the Correlation Coefficient

2. Select the cell into which you want to enter the **CORREL** function. In this example, we are going to place the function for the correlation in Cell B12.

3. Click on Cell B12 and type the CORREL function as follows:
   =CORREL(A2:A11,B2:B11)

   and press the Enter key,

   or

   use the Insert Formulas → More Functions → Statistical → CORREL using the Inserting a Function technique we talked about on pages 25 through 29 in Chapter 1 to enter the COR-REL function in Cell B12.

4. As you can see in Figure 5.2, the correlation coefficient was computed and the value returned to Cell B12. Notice that in the formula bar in Figure 5.2, you can see the CORREL function fully expressed and the value computed in Cell B12 as .692.

B12	▼	f_x	= CORREL(A2:A11,B2:B11)

	A	B	C	D
1	X	Y		
2	2	3		
3	4	2		
4	5	6		
5	6	5		
6	4	3		
7	7	6		
8	8	5		
9	5	4		
10	6	4		
11	7	5		
12		0.69213		

Figure 5.2 Computing the Correlation Coefficient Using the CORREL Function

Excel

PEARSON function is also a handy one to know. It returns the
of the Pearson product-moment correlation.

really much more convenient to name a range of values rather
than have to enter a particular set of cell references. So, instead of
A2:A11, you can just as well name the range correct, and then enter
that. You can review what you need to know about ranges in Appendix
A and in our discussion about ranges on pages 30 through 31.

A Visual Picture of a Correlation: The Scatterplot

There's a very simple way to visually represent a correlation:
Create what is called a **scatterplot,** or **scattergram.** This is simply a
plot of each set of scores on separate axes.

Here are the steps to complete a scattergram like you see in
Figure 5.3 for the 10 pairs of two scores for which we computed the
sample correlation above.

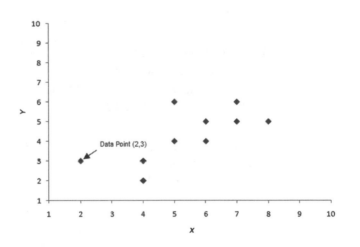

Figure 5.3 A Simple Scatterplot

1. Draw the x-axis and the y-axis. Usually, the X variable goes on the
horizontal axis and the Y variable goes on the vertical axis.

2. Mark both axes with the range of values that you know to be the case
for the data. For example, the value of the X variable in our example
ranges from 2 to 8, so we marked the x-axis from 0 to 9. There's no
harm in marking them a bit low or high—just as long as you allow
room for the values to appear. The value of the Y variable ranges from

2 to 6, and we marked that axis from 0 to 9. Having similarly labeled axes can sometimes make the finished scatterplot easier to understand.

3. Finally, for each pair of scores (such as 2 and 3, as shown in Figure 5.3), we entered a dot on the chart by marking the place where 2 falls on the x-axis and 3 falls on the y-axis. The dot represents a **data point,** which is the intersection of the two values, as you can see in Figure 5.3.

When all the data points are plotted, what does such an illustration tell us about the relationship between the variables? To begin with, the general shape of the collection of data points indicates whether the correlation is direct (positive) or indirect (negative).

A positive slope occurs when the data points group themselves in a cluster from the lower left-hand corner on the x- and y-axes through the upper right-hand corner. A negative slope occurs when the data points group themselves in a cluster from the upper left-hand corner on the x- and y-axes through the lower right-hand corner.

Here are some scatterplots showing very different correlations where you can see how the grouping of the data points reflects the sign and strength of the correlation coefficient.

Figure 5.4 shows a perfect direct correlation where $r_{xy} = 1.00$, and all the data points are aligned along a straight line with a positive slope.

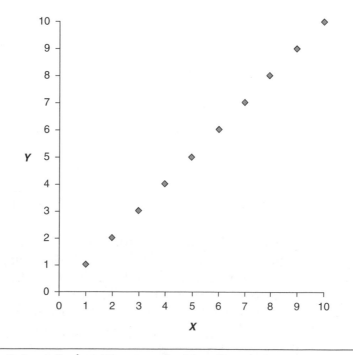

Figure 5.4 A Perfect Direct, or Positive, Correlation

If the correlation were perfectly indirect, the value of the correlation coefficient would be −1.0 and the data points would align themselves in a straight line as well, but from the upper left-hand corner of the chart to the lower right. In other words, the line that connects the data points would have a negative slope.

> Don't ever expect to find a perfect correlation between any two variables in the behavioral or social sciences. It would say that two variables are so perfectly correlated, they share everything in common. In other words, knowing one is like knowing the other. Just think about your classmates. Do you think they all share any one thing in common that is perfectly related to another of their characteristics across all these different people? Probably not. In fact, r values approaching .7 and .8 are just about the highest you'll see.

In Figure 5.5, you can see the scatterplot for a strong (but not perfect) direct relationship where $r_{xy} = .70$. Notice that the data points align themselves along a positive slope, although not perfectly.

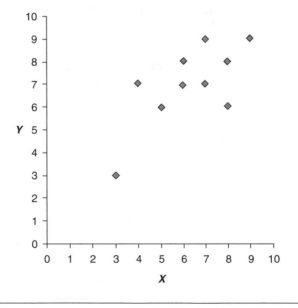

Figure 5.5   A Strong Positive, But Not Perfect, Direct Relationship

Now, we'll show you a strong indirect, or negative, relationship in Figure 5.6, where $r_{xy} = −.82$. Notice how the data points align themselves on a negative slope from the upper left-hand corner of the chart to the lower right-hand corner.

That's what different types of correlations look like, and you can really tell the general strength and direction by examining the way the points are grouped.

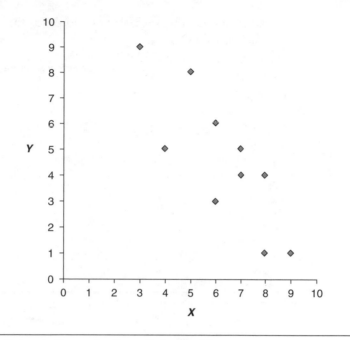

Figure 5.6 A Strong Indirect Relationship

Using Excel to Create a Scatterplot

Now we're talking. We spent a considerable amount of time and energy in Chapter 4 creating charts and then modifying certain elements.

One type of chart we did not review is a scatterplot (we left it for this chapter).

You already know what a scatterplot is, and here's exactly how you create one. We're still working with the data you first saw in Figure 5.1. In just a few clicks. . . . Amazing!

1. Highlight the entire range of values (not the column titles) from A2 through B11.

2. Click the Insert tab, and under Charts, click Scatter and the type of scatterplot you want to create. We clicked on the first one available, and here, you see it (with a little bit of fancy thrown in) in Figure 5.7.

Very cool indeed.

Figure 5.7 shows the fancy version of the scatterplot where we modified the scales and the colors used for the background and such, titled each axis, and gave the entire chart a title. But, this is absolutely a cinch to do—all it takes is some experimentation. Remember to save the worksheet as you work so you don't lose what you want and do lose what you don't!

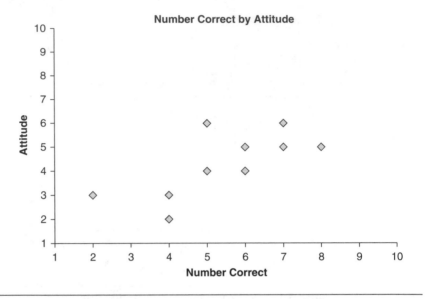

Figure 5.7   Create a Scatterplot in Three Magic Clicks

TECH
TALK
Not all correlations are reflected by a straight line showing the X and the Y values in a relationship called a **linear correlation**. The relationship may not be linear and may not be reflected by a straight line. Let's take the correlation between age and memory. For the early years, the correlation is probably highly positive—the older children get, the better their memories. Then, into young and middle adulthood, there isn't much of a change or much of a correlation, because most young and middle adults maintain a good memory. But with old age, memory begins to suffer, and there is an indirect relationship between memory and aging in the later years. If you take these together, you find that the correlation between memory and age tends to look something like a curve where memory increases, levels off, and then decreases. It's a curvilinear relationship, and sometimes, the best description of the relationship is a curvilinear one.

Bunches of Correlations: The Correlation Matrix

What happens if you have more than two variables? How are the correlations illustrated? Use a **correlation matrix** like the one shown below—a simple and elegant solution.

	Income	Education	Attitude	Vote
Income	1.00	0.35	−0.19	0.51
Education		1.00	−0.21	0.43
Attitude			1.00	0.55
Vote				1.00

As you can see, there are four variables in the matrix: level of income (Income), level of education (Education), attitude toward voting (Attitude), and level of participation (from 1 through 5) in the election or vote (Vote).

For each pair of variables, there is a correlation coefficient. For example, the correlation between income level and education is .35 (meaning that as income increases, so does level of education). The correlation between income level and attitude is −.19, meaning that income level and attitude are indirectly related.

In such a matrix where there are four variables, there are always $4!/(4 − 2)!2!$, or four things taken two at a time for a total of six correlation coefficients. Because variables correlate perfectly with themselves (those are the 1.00s down the diagonal), and because the correlation between Income and Vote is the same as the correlation between Vote and Income, the matrix creates a mirror image of itself.

You will see such matrices (the plural of matrix) when you read journal articles that use correlations to describe the relationship between several variables. You'll also see us generate a matrix when we use the Analysis ToolPak later on in this chapter to deal with the correlation between more than two variables.

MORE EXCEL BUNCHES OF CORRELATIONS À LA EXCEL

If you want to get ambitious and increase the power of Excel for you, you can create a matrix like the kind we showed above by placing the appropriate form of the CORREL function in the appropriate cell, as you see in Figure 5.8. Here, we used both the name range option on the Insert menu (remember about naming ranges from Appendix A) and the CORREL function.

But—and here's the big but—we are showing you the formulas in the cells rather than the results. How did we do this? You can see the formulas and functions, rather than the results that the formulas and functions return, by highlighting the entire spreadsheet (Ctrl+A), and then using the CTRL+` key combination, which toggles between the results of the formulas and functions and the formulas and functions themselves. The data for the creation of the matrix appear in four columns, each representing one of four variables (Income, Education, Attitude, and Vote), and the functions for computing the correlations in the matrix appear below the columns.

	A	B	C	D	E
1		Income	Education	Attitude	Vote
2		74190	13	1	1
3		80931	12	3	2
4		81314	11	4	2
5		73089	11	5	2
6		62023	11	3	2
7		61217	10	4	2
8		84526	11	5	1
9		87251	11	4	1
10		62659	12	5	2
11		76450	10	6	2
12		70512	12	7	2
13		78858	9	6	1
14		78628	13	7	1
15		86212	14	8	2
16		74962	9	8	2
17		58828	11	9	4
18		61471	10	8	5
19		78621	12	7	5
20		60071	9	8	4
21					
22		Income	Education	Attitude	Vote
23	Income	=CORREL(Income,Income)	=CORREL(Income,Education)	=CORREL(Income,Attitude)	=CORREL(Income,Vote)
24	Education		=CORREL(Education,Education)	=CORREL(Attitude,Education)	=CORREL(Vote,Education)
25	Attitude			=CORREL(Attitude,Attitude)	=CORREL(Attitude,Vote)
26	Vote				=CORREL(Vote,Vote)
27					

Figure 5.8   Showing the Use of the Ctrl+` Key Combination to Reveal the CORREL Function in Various Cells to Create a Correlation Matrix

USING THE AMAZING ANALYSIS TOOLPAK TO COMPUTE CORRELATIONS

If you think that the CORREL function is amazing, you "ain't seen nothin' yet" (thanks to Humphrey Bogart). Although the CORREL function is very useful, the Correlation tool in the Analysis ToolPak is even easier to use.

Let's return to the data that we see in Figure 5.8 and follow these steps.

1. Click the Data tab, and then click Data Analysis. You will see the Data Analysis dialog box. Need a brush up on how to use the ToolPak? See Little Chapter 1b.

2. Click Correlation, and then click OK, and you will see the Correlation dialog box as shown in Figure 5.9.

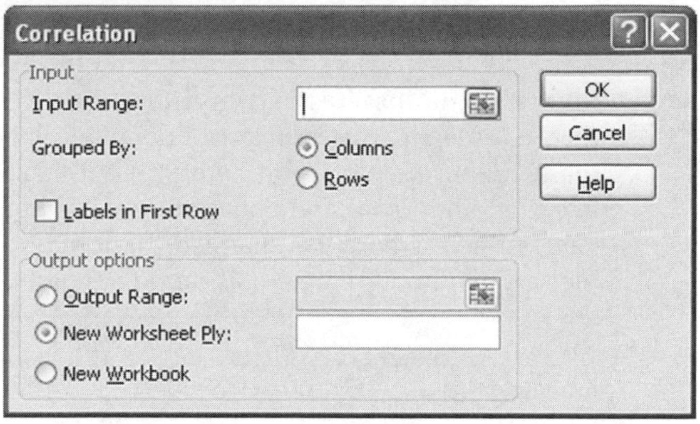

Figure 5.9 The Correlation Dialog Box

3. In the Input Range box, enter the range of data you want Excel to use in the computation of the correlations. Be sure to include the column headings (so the headings show up in the analysis). As you can see in Figure 5.10, the data we want to analyze are in Cells B1 through E20.

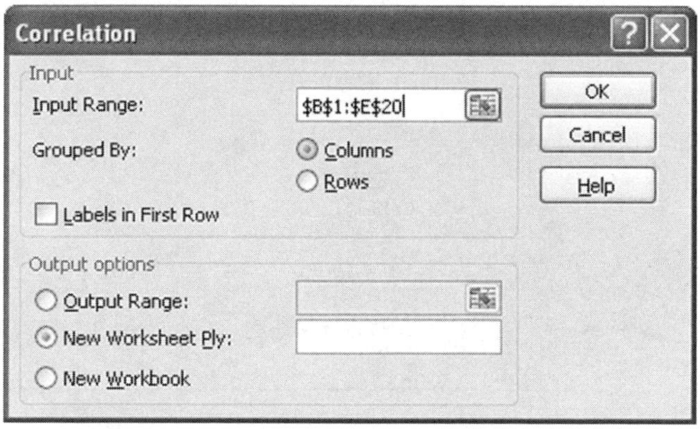

Figure 5.10 Entering the Input Range Information in the Correlation Dialog Box

4. Click Labels in the First Row check box.

5. Now click the Output Range button in the Output options section of the dialog box, and enter the location where you want Excel to return the results of the analysis. In this example, we checked $B$22.

6. Click OK, and there it is, folks, in Figure 5.11, only done with very little effort on your part. Truly a time to rejoice.

More Excel

Keep in mind that it is Excel's decision to create the matrix in the format you see with the number of decimal places in Figure 5.11 (far too many, in our opinion) and with the font that is used. These entries are like any other data, and you can do with them what you want, as you can see here where we changed the format to be a bit more intelligible, rounded off the decimals, and changed the general appearance.

	Income	Education	Attitude	Age
Income	1.00			
Education	0.35	1.00		
Attitude	−0.19	−0.21	1.00	
Age	−0.51	−0.20	0.55	1.00

When a tool from the ToolPak is used, it returns a value rather than the result of a formula, so it is a number that is in that cell, and not a formula or a function. You can do just about anything you want with it.

	A	B	C	D	E	F
1		Income	Education	Attitude	Vote	
2		$ 74,190	13	1	1	
3		$ 80,931	12	3	2	
4		$ 81,314	11	4	2	
5		$ 73,089	11	5	2	
6		$ 62,023	11	3	2	
7		$ 61,217	10	4	2	
8		$ 84,526	11	5	1	
9		$ 87,251	11	4	1	
10		$ 62,659	12	5	2	
11		$ 76,450	10	6	2	
12		$ 70,512	12	7	2	
13		$ 78,858	9	6	1	
14		$ 78,628	13	7	1	
15		$ 86,212	14	8	2	
16		$ 74,962	9	8	2	
17		$ 58,828	11	9	4	
18		$ 61,471	10	8	5	
19		$ 78,621	12	7	5	
20		$ 60,071	9	8	4	
21						
22			Income	Education	Attitude	Vote
23		Income	1			
24		Education	0.346054385	1		
25		Attitude	-0.190313082	-0.207210094	1	
26		Vote	-0.507643545	-0.200229322	0.551954588	1
27						

Figure 5.11   Using the Correlation Analysis ToolPak to Create a Matrix of Correlations

UNDERSTANDING WHAT THE CORRELATION COEFFICIENT MEANS

Well, we have this numerical index of the relationship between two variables, and we know that the higher the value of the correlation (regardless of its sign), the stronger the relationship is. But because the correlation coefficient is a value that is not directly tied to the value of an outcome, just how can we interpret it and make it a more meaningful indicator of a relationship?

Here are different ways to look at the interpretation of that simple r_{xy}.

Using-Your-Thumb Rule

Perhaps the easiest (but not the most informative) way to interpret the value of a correlation coefficient is by eyeballing it and using the information in Table 5.2.

Table 5.2 Interpreting a Correlation Coefficient

Size of the Correlation	Coefficient General Interpretation
.8 to 1.0	Very strong relationship
.6 to .8	Strong relationship
.4 to .6	Moderate relationship
.2 to .4	Weak relationship
.0 to .2	Weak or no relationship

So, if the correlation between two variables is .5, you could safely conclude that the relationship is a moderate one—not strong, but certainly not weak enough to say that the variables in question don't share anything in common.

This eyeball method is perfectly acceptable for a quick assessment of the strength of the relationship between variables, such as a description in a research report. But because this rule of thumb does depend on a subjective judgment (of what's "strong" or "weak"), we would like a more precise method. That's what we'll look at now.

And there is some overlap in Table 5.2 such that a correlation of .4, for example, can fall into the weak or moderate relationship range. It's your call—further emphasizing how numbers don't tell the whole story—you do.

A Determined Effort: Squaring the Correlation Coefficient

Here's the much more precise way to interpret the correlation coefficient: computing the coefficient of determination. The **coefficient**

of determination is the percentage of variance in one variable that is accounted for by the variance in the other variable. Quite a mouthful, huh?

Earlier in this chapter, we pointed out how variables that share something in common tend to be correlated with one another. If we correlated math and English grades for 100 fifth-grade students, we would find the correlation to be moderately strong, because many of the reasons why children do well (or poorly) in math tend to be the same reasons why they do well (or poorly) in English. The number of hours they study, how bright they are, how interested their parents are in their schoolwork, the number of books they have at home, and more are all related to both math and English performance and account for differences between children (and that's where the variability comes in).

The more these two variables share in common, the more they will be related. These two variables share variability—or the reason why children differ from one another. And on the whole, the brighter child who studies more will do better. To determine exactly how much of the variance in one variable can be accounted for by the variance in another variable, the coefficient of determination is computed by squaring the correlation coefficient.

For example, if the correlation between GPA and number of hours of study time is .70 (or $r_{GPA \cdot time} = .70$), then the coefficient of determination, represented by r^2, is $.7^2$, or .49. This means that 49% of the variance in GPA can be explained by the variance in studying time. And the stronger the correlation, the more variance can be explained (which only makes good sense). The more two variables share in common (such as good study habits, knowledge of what's expected in class, and lack of fatigue), the more information about performance on one score can be explained by the other score.

However, if 49% of the variance can be explained, this means that 51% cannot—so even for a strong correlation of .70, many of the reasons why scores on these variables tend to be different from one another go unexplained. This amount of unexplained variance is called the **coefficient of alienation** (also called the **coefficient of non-determination**). Don't worry. No aliens here. This isn't *X-Files* stuff, it's just the amount of variance in *Y* not explained by *X* (and, of course, vice versa).

How about a visual presentation of this sharing variance idea? OK. In Figure 5.12, you'll find a correlation coefficient, the corresponding coefficient of alienation, and a diagram that represents how much variance is shared between the two variables. The larger the striped area in each diagram (and the more variance the two variables share), the more highly the variables are correlated.

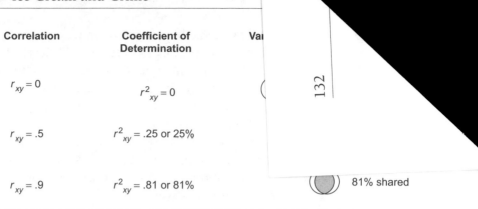

Correlation	Coefficient of Determination	Va
$r_{xy} = 0$	$r^2_{xy} = 0$	132
$r_{xy} = .5$	$r^2_{xy} = .25$ or 25%	
$r_{xy} = .9$	$r^2_{xy} = .81$ or 81%	81% shared

Figure 5.12 How Variables Share Variance and the Resulting Correlation

The first diagram shows two circles that do not touch. They don't touch because they do not share anything in common. The correlation is 0.

The second diagram shows two circles that overlap. With a correlation of .5 (and r^2_{xy} = .25), they share about 25% of the variance between themselves.

Finally, the third diagram shows that the two circles are almost placed one on top of the other. With an almost perfect correlation of r_{xy} = .90 (r^2_{xy} = .81), they share about 81% of the variance between themselves.

AS MORE ICE CREAM IS EATEN . . . THE CRIME RATE GOES UP (OR ASSOCIATION VERSUS CAUSALITY)

Now, here's the really important thing to be careful about when computing, reading about, or interpreting correlation coefficients. Imagine this. In a small Midwestern town, a phenomenon was discovered that defied any logic. The local police chief observes that as ice cream consumption increases, crime rates tend to increase as well. Quite simply, if you measured both, you would find the relationship was direct, which means that as people eat more ice cream, the crime rate increases. And as you might expect, as they ate less ice cream, the crime rate went down. The police chief was baffled until he recalled the Stat 1 class he took in college and still fondly remembered.

His wondering about how this could be true turned into an Aha! Moment; "Very easily," he thought. The two variables must share something or have something in common with one another. Remember that it must be something that relates to both level of ice cream consumption and level of crime rate. Can you guess what that is?

The outside temperature is what they both have in common. When it gets warm outside, such as in the summertime, more crimes are committed (it stays light longer, people leave the windows open, etc.). And because it is warmer, people enjoy the ancient treat and art of eating ice cream. And conversely, during the long and dark winter months, less ice cream is consumed and fewer crimes are committed as well.

Joe Bob, recently elected as a city commissioner, learns about these findings and has a great idea, or at least one that he thinks his constituents will love. (Keep in mind, he skipped the statistics offering in college.) Why not just limit the consumption of ice cream in the summer months, which will surely result in a decrease in the crime rate? Sounds good, right? Well, on closer inspection, it really makes no sense at all.

That's because of the simple principle that *correlations express the association that exists between two or more variables; they have nothing to do with causality.* In other words, just because level of ice cream consumption and crime rate increase together (and decrease together as well) does not mean that a change in one necessarily results in a change in the other.

For example, if we took all the ice cream out of all the stores in town and no more was available, do you think the crime rate would decrease? Of course not, and it's preposterous to think so. But strangely enough, that's often how associations are interpreted—as being causal in nature—and complex issues in the social and behavioral sciences are reduced to trivialities because of this misunderstanding. Did long hair and hippiedom have anything to do with the Vietnam conflict? Of course not. Does the rise in the number of crimes committed have anything to do with more efficient and safer cars? Of course not. But they all happened (and happen) at the same time, which creates the illusion of being associated.

OTHER COOL CORRELATIONS

There are different ways in which variables can be assessed. For example, nominal-level variables are categorical in nature, such as race (black or white) or political affiliation (Independent or Republican). Or, if you are measuring income and age, these are both interval-level variables, because the underlying continuum on which they are based has equally appearing intervals. As you continue your studies, you're likely to come across correlations between data that occur at different levels of measurement. And to compute these correlations, you need some specialized techniques. Table 5.3 summarizes what these different techniques are and how they differ from one another.

TABLE 5.3 Correlation Coefficient Shopping, Anyone?

| Level of Measurement and Examples | | Type of Correlation | Correlation Being Computed |
Variable X	Variable Y		
Nominal (voting preference, such as Republican or Democrat)	Nominal (sex, such as male or female)	Phi coefficient	The correlation between voting preference and sex
Nominal (social class, such as high, medium, or low)	Ordinal (rank in high school graduating class)	Rank biserial coefficient	The correlation between social class and rank in high school
Nominal (family configuration, such as intact or single parent)	Interval (grade point average)	Point biserial	The correlation between family configuration and grade point average
Ordinal (height converted to rank)	Ordinal (weight converted to rank)	Spearman rank coefficient	The correlation between height and weight
Interval (number of problems solved)	Interval (age in years)	Pearson correlation coefficient	The correlation between number of problems solved and age in years

Summary

The idea of showing how things are related to one another and what they have in common is a very powerful idea and a very useful descriptive statistic (used in inference as well). Keep in mind that correlations express a relationship that is only associative and not causal, and you'll be able to understand how this statistic gives us valuable information about relationships and how variables change or remain the same in concert with others.

Time to Practice

1. Use these data to answer Questions 1a and 1b. These data are saved as Chapter 5 Data Set 1.

Total No. of Problems Correct (out of a possible 20)	Attitude Toward Test Taking (out of a possible 100)
17	94
13	73
12	59
15	80
16	93
14	85
16	66
16	79
18	77
19	91

a. Compute the Pearson product-moment correlation coefficient by hand and show all your work.
b. Construct a scatterplot for these 10 values by hand. Based on the scatterplot, would you predict the correlation to be direct or indirect? Why?

2. Use these data to answer Questions 2a and 2b.

Speed in Seconds (to complete a 50-yard swim)	Strength (no. of pounds bench-pressed)
21.6	135
23.4	213
26.5	243
25.5	167
20.8	120
19.5	134
20.9	209
18.7	176
29.8	156
28.7	177

a. Using either a calculator or Excel, compute the Pearson correlation coefficient.
b. Interpret these data using the general range of very weak to very strong, and also compute the coefficient of determination. How does the subjective analysis compare to the value of r^2?

3. Compute the correlation coefficient for the following data and interpret it. It's available as Chapter 5 Data Set 2. We are examining the number of years of training that doctors have and their success using a certain procedure with the doctors being evaluated on a scale from 1 to 10 with 1 being ☺ and 10 being ☺☺☺. Here's a tip— remember that it is good to have a low score on the second variable.

4. The coefficient of determination between two variables is .64.

   Answer the following questions:

   a. What is the Pearson correlation coefficient?
   b. How strong is the relationship?
   c. How much of the variance in the relationship between these two variables is unaccounted for?

5. Look at Table 5.3. What type of correlation coefficient would you use to examine the relationship between ethnicity (defined as different categories) and political affiliation (defined as Democrat, Republican, or unaffiliated)? How about club membership (yes or no) and high school GPA? Explain why you selected the answers you did.

6. You're the director of the local health clinic, and you are working on a grant with the school district to examine the relationship between dropout rate and teenage pregnancy. One of the local legislators has gotten a copy of the study and made the following statement to the press, "It's obvious that the longer children stay in school, the less likely they are to become pregnant." What's wrong with that statement?

Answers to Practice Questions

1a. $r = .596$.

1b. From the answer to 1a, you already know that the correlation is direct. But from the scatterplot shown in Figure 5.13, you can predict it to be such (without actually knowing the sign of the coefficient), because the data points group themselves from the lower left corner of the graph to the upper right corner and assume a positive slope.

2a. $r = .269$.

2b. According to the table presented earlier in the chapter, the general strength of the correlation of this magnitude is weak. The coefficient of determination is $.269^2$ or .072 (or 7.2%) of the variance is accounted for. The subjective analysis (weak) and the objective one (7.2% of the variance accounted for) are consistent with one another.

3. The correlation is .45, which means the more training the doctors have, the better the outcomes (remember that 1 is good and 10 is not) and that the less training the doctors have, the worse the outcomes. Squaring the .45 results in a value of .20, or 20% of the

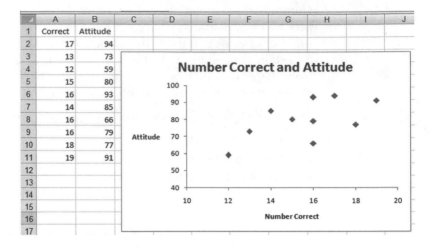

	A	B	C	D	E	F	G	H	I	J
1	Correct	Attitude								
2	17	94								
3	13	73								
4	12	59								
5	15	80								
6	16	93								
7	14	85								
8	16	66								
9	16	79								
10	18	77								
11	19	91								
12										
13										
14										
15										
16										
17										

Figure 5.13   A Scatterplot of Two Variables That Are Positively Related

variability in one variable being accounted for by the other. In real life, it does not seem that the number of procedures a doctor has done in his or her training and the outcomes really share a lot.

4a. .8.

4b. Very strong.

4c. 1 − .64, or 36% (.36).

5. To examine the relationship between ethnicity and political affiliation, you would use the phi coefficient because both variables are nominal in nature. To examine the relationship between club membership and high school GPA, you would use the point biserial correlation because one variable is nominal (club membership), and the other is interval (GPA).

6. Well, the legislator may be very well intentioned, but he or she may also be somewhat wrong. The statement implies that teenage pregnancy may be caused by dropping out of school (and we sure know that's not the case ☺). In fact, these two variables are just related to one another, and the relationship might very well be a function of what they share in common (such as family income level and other important variables related to both early pregnancy and dropout rates).

6 Just the Truth

An Introduction to Understanding Reliability and Validity

Difficulty Scale ☺☺☺ (not so hard)

How Much Excel? 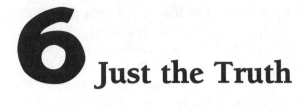 (just a mention)

What you'll learn about in this chapter

- What reliability and validity are and why they are important
- This is a stat class! What's up with this measurement stuff?
- The basic measurement scales
- How to compute and interpret various types of reliability coefficients
- How to compute and interpret various types of validity coefficients

AN INTRODUCTION TO RELIABILITY AND VALIDITY

Professionals in the field of social welfare (and in other fields) recognize that the existence of more than a half million foster children in the United States is a serious concern. One of the major issues is how foster children adjust to their temporary adoptive families given that their biological families still play a very important role in their lives.

Sonya J. Leathers examined this question when she studied whether frequent parental visiting was associated with foster children's allegiances to foster families and biological parents. In a sample of 199 adolescents, she found that frequent visits to their family of origin did create conflicts, and she suggested interventions that might help those conflicts be minimized.

To complete her study, she used a variety of different dependent variables (such as the Children's Symptom Inventory as well as interviews). Among other things, what she really got right is her care in selecting measurement instruments that had established, and acceptable, levels of reliability and validity—not a step that every researcher takes and one that we focus on in this chapter.

Want to know more? Check out the original reference: Leathers, S. (2003). Parental visiting, conflicting allegiances, and emotional and behavioral problems among foster children. *Family Relations, 52,* 53–63.

What's Up With This Measurement Stuff?

A very good question. After all, you enrolled in a stat class, and up to now, that's been the focus on the material that has been covered. Now it looks like you're faced with a topic that belongs in a test and measurement class. So, what's this material doing in a stat book?

An excellent question, and one that you should be asking. Why? Well, most of what we have covered so far in *Statistics for People Who (Think They) Hate Statistics . . . Excel 2007 Edition* has to do with the collection, analysis, and interpretation of data. A very significant part of those activities is the collection part, and a significant part of collecting data is making sure that the data are what you think they are—that the data represent what it is you want to know about. In other words, if you're studying poverty, you want to make sure that the measure you use to assess poverty works. Or, if you are studying aggression in middle-aged males, you want to make sure that whatever tool you use to assess aggression, works.

More really good news: Should you continue in your education and want to take a class on tests and measurement, this introductory chapter will give you a real jump on understanding the scope of the area and what topics you'll be studying.

And in order to make sure that the entire process of collecting data and making sense out of them works, you first have to make sure that what you use to collect data works as well. The fundamental questions that will be answered in this chapter are, "How do I know that the test, scale, instrument, etc., I use works every time I use it?" (that's reliability) and "How do I know that the test, scale, instrument, etc., I use measures what it is supposed to?" (that's validity).

More Excel

In Chapter 5, we did a pretty extensive job of introducing correlation coefficients and how they are used, and we'll also be talking about them later in this chapter (and more in Chapter 14). Whatever you have already learned about using Excel is directly applicable here. Excel does not have tools named something like "reliability calculator" as such, but instead uses the CORREL function and the Correlation tool from the Analysis ToolPak.

TECH TALK Anyone who does research will tell you of the importance of establishing the reliability and validity of your test tool, whether it's a simple observational instrument of consumer behavior or one that measures a complex psychological construct such as attachment. However, there's another very good reason. If the tools that you use to collect data are unreliable or invalid, then the results of any test of any hypothesis have to be inconclusive. If you are not sure that the test does what it is supposed to and that it does so consistently, how do you know that the results you got are a function of the lousy test tools rather than fair and honest results of the hypothesis you are testing in the first place? Want a clean test of the hypothesis? Make reliability and validity your business.

ALL ABOUT MEASUREMENT SCALES

Before we can talk much about reliability and validity, we first have to talk about different **scales of measurement.** What is **measurement**? The assignment of values to outcomes following a set of rules—simple. The results are the different scales we'll define in a moment, and an outcome is anything we are interested in measuring, such as hair color, gender, test score, or height.

These scales of measurement, or rules, are particular levels at which outcomes are measured. Each level has a particular set of characteristics. And scales of measurement come in four flavors (there are four types): nominal, ordinal, interval, and ratio. Let's move on to a brief discussion and examples of the four scales of measurement.

A Rose by Any Other Name:
The Nominal Level of Measurement

The **nominal level of measurement** is defined by the characteristics of an outcome that fits into one and only one class or category. For example, gender can be a nominal variable (female and male), as can ethnicity (Caucasian or African American), as can political affiliation (Republican, Democrat, or Independent). Nominal-level variables are "names" (nominal in Latin), and this is the least precise level of measurement. Nominal levels of measurement have categories that are mutually exclusive; for example, political affiliation cannot be both Republican and Democrat.

Any Order Is Fine With Me:
The Ordinal Level of Measurement

The "ord" in **ordinal level of measurement** stands for order, and the characteristic of things being measured here is that they are ordered. The perfect example is a rank of candidates for a job. If we know that Russ is ranked #1, Sheldon is ranked #2, and Hannah is ranked #3, then this is an ordinal arrangement. We have no idea how much higher on this scale Russ is relative to Sheldon than Sheldon is relative to Hannah. We just know that it's better to be #1 than #2 than #3, but not by how much.

1 + 1 = 2: The Interval Level of Measurement

Now we're getting somewhere. When we talk about the **interval level of measurement,** it is where a test or an assessment tool is based on some underlying continuum such that we can talk about how much more a higher performance is than a lesser one. For example, if you get 10 words correct on a vocabulary test, that is twice as many as getting five words correct. A distinguishing characteristic of interval-level scales is that the intervals along the scale are equal to one another. Ten words correct is two more than eight correct, which is three more than five correct.

Can Anyone Have Nothing of Anything?
The Ratio Level of Measurement

Well, here's a little conundrum for you. An assessment tool at the **ratio level of measurement** is characterized by the presence of an absolute zero on the scale. What that means is the absence of any of the trait that is being measured. The conundrum? Are there outcomes we measure where it is possible to have nothing of what is being measured? In some disciplines, that can be the case. For example, in the physical and biological sciences, you can have the absence of a characteristic, such as absolute zero (no molecular movement) or zero light. You can even be weightless in space, meaning that you don't weigh anything, or a scale can be ratio because it can record a reading of zero. In the social and behavioral sciences, it's a bit harder. Even if you score 0 on that spelling test or miss every item of an IQ test (in Russian), it does not mean that you have no spelling ability or no intelligence, right?

In Sum . . .

These scales of measurement, or rules, represent particular levels at which outcomes are measured. And, in sum, we can say the following:

- Any outcome can be assigned to one of the four scales of measurement we discussed above.
- Scales of measurement have an order, from the least precise being nominal, to the most precise being ratio.
- The "higher up" the scale of measurement, the more precise the data being collected, and the more detailed and informative the data are. It may be enough to know that some people are rich and some poor (and that's a nominal or categorical distinction), but it's much better to know exactly how much money one makes (interval or ratio). We can always make the "rich"/"poor" distinction if we want to once we have all the information.
- Finally, the more precise scales (such as interval) contain all the qualities of the scales below it, including ordinal and nominal. If you know that the Bears' batting average is .350, you know it is better than the Tigers' batting average (.250) by 100 points, but you also know that the Bears are better than

the Tigers (but not by how much), and that the Bears are different from the Tigers (but there's no direction to the difference).

And in sum again, take a look at this table, which shows you what you need to conclude what the correct level of measurement is.

	Characteristics			
Scale	Absolute Zero	Equidistant Points	Ranked Data	Data in Categorical
Ratio	✓	✓	✓	✓
Interval		✓	✓	✓
Ordinal			✓	✓
Nominal				✓

RELIABILITY—DOING IT AGAIN UNTIL YOU GET IT RIGHT

Reliability is pretty easy to figure out. It's simply whether a test, or whatever you use as a measurement tool, measures something consistently. If you administer a test of personality before a special treatment occurs, will the administration of that same test 4 months later be reliable? That, my friend, is one of the questions. And that is why there are different types of reliability, each of which we will get to after we define reliability just a bit more.

Test Scores—Truth or Dare

When you take a test in this class, you get a score, such as 89 (good for you) or 65 (back to the book!). That test score consists of several different elements, including the **observed score** (or what you actually get on the test, such as 89 or 65) and a **true score** (the true, 100% accurate reflection of what you really know). We can't directly measure true score because it is a theoretical reflection of the actual amount of the trait or characteristic possessed by the individual.

TECH TALK Nothing about this tests and measurement stuff is clear cut, and this true score stuff surely qualifies. Here's why. We just defined true score as the real, real, real value associated with some trait or attribute. So far so good. But there's another point of view as well. Some psychometricians (the people who do tests and measurement for a living) believe that true score has nothing to do with whether the construct of interest is really being reflected. Rather, true score is the *mean* score an individual would get if he or she took a test an infinite number of times, and it represents the theoretical typical level of performance on a given test. Now, one would hope that the typical level of performance would reflect the construct of interest, but that's another question (a validity one at that). The distinction here is that a test is reliable if it consistently produces whatever score a person would get on average, regardless of whatever it is the test is measuring. In fact, a perfectly reliable test might not produce a score that has anything to do with the construct of interest, such as "what you really know."

Why aren't true and observed scores the same? Well, they can be if the test (and the accompanying observed score) is a perfect (and we mean absolutely perfect) reflection of what's being measured.

But the Yankees don't always win, the bread sometimes falls on the buttered side, and Murphy's Law tells us that the world is not perfect. So, what you see as an observed score may come close to the true score, but rarely are they the same. Rather, the difference as you see here is in the amount of error that is introduced.

Observed Score = True Score + Error Score

Error? Yes—in all its glory. For example, let's suppose for a moment that someone gets an 80 on a stat test, but his or her true score (which we never really know but can only theorize) is 89. That nine-point difference (that's the **error score**) is due to error, or the reason why individual test scores vary from being 100% true.

What might be the source of such error? Well, perhaps the room in which the test is taken is so warm that it causes you to fall asleep. That would certainly have an impact on your test score. Or, perhaps you didn't study for the test as much as you should have. Ditto. Both of these examples would reflect testing situations or conditions rather than qualities of the trait being measured, right? Our job is to reduce those errors as much as possible by having, for example, good test-taking conditions and making sure you are encouraged to get enough sleep. Reduce the error and you increase the reliability, because the observed score more closely matches the true score.

The less error, the more reliable—it's that simple.

Different Types of Reliability

There are several different types of reliability, and we'll cover the four most important and most often used in this section. They are all summarized in Table 6.1.

Table 6.1   Different Types of Reliability, When They Are Used, How They Are Computed, and What They Mean

Type of Reliability	When You Use It	How You Do It	An Example of What You Can Say When You're Done
Test–retest reliability	When you want to know whether a test is reliable over time	Correlate the scores from a test given in Time 1 with the same test given in Time 2.	The Bonzo test of identity formation for adolescents is reliable over time.
Parallel forms reliability	When you want to know if several different forms of a test are reliable or equivalent	Correlate the scores from one form of the test with the scores from a second form of the same test of the same content (but not the exact same test).	The two forms of the Regular Guy test are equivalent to one another and have shown parallel forms reliability.
Internal consistency reliability	When you want to know if the items on a test assess one, and only one, dimension	Correlate each individual item score with the total score.	All of the items on the SMART Test of Creativity assess the same construct.
Interrater reliability	When you want to know whether there is consistency in the rating of some outcome	Examine the percent of agreement between raters.	The interrater reliability for the best-dressed Foosball player judging was .91, which indicates a high degree of agreement between judges.

TEST–RETEST RELIABILITY

Test–retest reliability is used when you want to examine whether a test is reliable over time. For example, let's say that you are developing a test that will examine preferences for different types of vocational programs.

You may administer the test in September and then readminister the same test (and it's important that it be the same) again in June. Then, the two sets of scores (remember, the same people took it twice) are correlated, and you have a measure of reliability. Test–retest reliability is a must when you are examining differences or changes over time.

You must be very confident that what you are measuring has been measured in a reliable way such that the results you are getting come as close as possible to the individual's score each and every time.

Computing Test–Retest Reliability. Here are some scores from a test at Time 1 and Time 2 for the MVE (Mastering Vocational Education Test) under development. Our goal is to compute the Pearson correlation coefficient as a measure of the test–retest reliability of the instrument.

ID	Score From Test 1	Score From Test 2
1	54	56
2	67	77
3	67	87
4	83	89
5	87	89
6	89	90
7	84	87
8	90	92
9	98	99
10	65	76

The first and last step in this process is to compute the Pearson product-moment correlation (see Chapter 5 for a refresher on using the CORREL function and the correlation tool in the Analysis ToolPak), which is equal to

$$r_{\text{Time1} \cdot \text{Time2}} = .90. \qquad (6.1)$$

We'll get to the interpretation of this value shortly.

PARALLEL FORMS RELIABILITY

Parallel forms reliability is used when you want to examine the equivalence or similarity between two different forms of the same test.

For example, let's say that you are doing a study on memory, and part of the task is to look at 10 different words, memorize them as best as you can, and then recite them after 20 seconds of study and 10 seconds of rest. Because this is a study that takes place over a 2-day period and involves some training of memory skills, you want to have another set of items that is exactly similar in task demands, but obviously cannot be the same as far as content. So, you create another list of words that is hopefully similar to the first. In this example, you want the consistency to be high across forms—the same ideas being tested, just using a different form.

Computing Parallel Forms Reliability. Here are some scores from the IRMT (I Remember Memory Test) on Form A and Form B. Our goal is to compute the Pearson correlation coefficient as a measure of the parallel forms reliability of the instrument.

ID	Scores From Form A	Scores From Form B
1	4	5
2	5	6
3	3	5
4	6	6
5	7	7
6	5	6
7	6	7
8	4	8
9	3	7
10	3	7

The first and last step in this process is to compute the Pearson product-moment correlation (see Chapter 5 for a refresher on this), which is equal to

$$r_{\text{FormA} \cdot \text{FormB}} = .13. \tag{6.2}$$

We'll get to the interpretation of this value shortly.

INTERNAL CONSISTENCY RELIABILITY

Internal consistency reliability is quite different from the two previous types that we have explored. It is used when you want to know whether the items on a test are consistent with one another in that they represent one, and only one, dimension, construct, or area of interest.

Let's say that you are developing a test of attitudes toward different types of health care, and you want to make sure that the set of five items measures just that, and nothing else. You would look at the score for each item (for a group of test takers) and see if the individual score correlates with the total score. You would expect that people who scored high on certain items (e.g., "I like my HMO") would have scored low on others (e.g., "I don't like spending money on health care"), and that this would be consistent across all the people who took the test.

Computing Cronbach's Alpha (or α). Here are some sample data for 10 people on this five-item attitude test (the I♥HMO test) where scores are between 1 (*strongly disagree*) and 5 (*strongly agree*) on each item.

TECH TALK When you compute Cronbach's alpha (named after Lee Cronbach), you are actually correlating the score for each item with the total score for each individual, and comparing that to the variability present for all individual item scores. The logic is that any individual test taker with a high total test score should have a high(er) score on each item (such as 5, 5, 3, 5, 3, 4, 4, 2, 4, 5) for a total score of 40, and that any individual test taker with a low(er) total test score should have a low(er) score on each individual item (such as 5, 1, 5, 1, 5, 5, 1, 5, 5, 1, 5, 1) for a total score of 40 as well, but much less unified or one-dimensional.

ID	Item 1	Item 2	Item 3	Item 4	Item 5
1	3	5	1	4	1
2	4	4	3	5	3
3	3	4	4	4	4
4	3	3	5	2	1
5	3	4	5	4	3
6	4	5	5	3	2
7	2	5	5	3	4
8	3	4	4	2	4
9	3	5	4	4	3
10	3	3	2	3	2

And, here's the formula to compute Cronbach's alpha:

$$\alpha = \left(\frac{k}{k-1}\right)\left(\frac{s_y^2 - \Sigma s_i^2}{s_y^2}\right),$$ (6.3)

where

k = the number of items,

s_y^2 = the variance associated with the observed score, and

Σs_i^2 = the sum of all the variances for each item.

Here's the same set of data with the values (the variance associated with the observed score, or s_y^2, and the sum of all the variances for each item) needed to complete the above equation, or Σs_i^2.

ID	Item 1	Item 2	Item 3	Item 4	Item 5	Total Score
1	3	5	1	4	1	14
2	4	4	3	5	3	19
3	3	4	4	4	4	19
4	3	3	5	2	1	14
5	3	4	5	4	3	19
6	4	5	5	3	2	19
7	2	5	5	3	4	19
8	3	4	4	2	4	17
9	3	5	4	4	3	19
10	3	3	2	3	2	13
						$s_y^2 = 6.4$
Item Variance	0.32	0.62	1.96	0.93	1.34	$\Sigma s_i^2 = 5.17$

And when you plug all these figures into the equation and get the following equation,

$$\alpha = \left(\frac{5}{5-1}\right)\left(\frac{6.40 - 5.17}{6.4}\right) = .24,$$ (6.4)

you find that coefficient alpha is .24, and you're done (except for the interpretation that comes later!). Sorry—no cool Excel tools here for us to use.

TECH TALK If we told you that there were many other types of internal consistency reliability, you would not be surprised, right? This is especially true for measures of internal consistency. Not only is there coefficient alpha, but also split-half reliability, Spearman-Brown, Kuder-Richardson 20 and 21 (KR_{20} and KR_{21}), and others that basically do the same thing—examine the one-dimensional nature of a test—only in different ways.

INTERRATER RELIABILITY

Interrater reliability is the measure that tells you how much two raters agree on their judgments of some outcome.

For example, let's say you are interested in a particular type of social interaction during a transaction between a banker and a potential checking account customer, and you observe both people in real time (you're observing behind a one-way mirror) to see if the new and improved customer relations course that the banker took resulted in increased smiling and pleasant types of behavior toward the potential customer. Your job is to note every 10 seconds if the banker is demonstrating one of the three different behaviors he has been taught—smiling, leaning forward in his chair, or using his hands to make a point. Each time you see any one of those behaviors, you mark it on your scoring sheet as a slash (/). If you observe nothing, you score a dash (–).

As part of this process, and to be sure that what you are recording is a reliable measure, you will want to find out what the level of agreement is between observers as to the occurrence of these behaviors. The more similar the ratings are, the higher the level of interrater agreement and interrater reliability.

Computing Interrater Reliability. In this example, the really important variable here is whether or not a customer-friendly act occurred within a set of 10-second time frames across 2 minutes (or twelve 10-second periods). So, what we are looking at is the rating consistency across a 2-minute period broken down into twelve 10-second periods. A slash (/) on the scoring sheet means that the behavior occurred and a dash (–) means it did not.

	Period →	1	2	3	4	5	6	7	8	9	10	11	12
Rater 1	Dave	/	–	/	/	/	–	/	/	–	–	/	/
Rater 2	Maureen	/	–	/	/	/	–	/	/	–	/	–	/

For a total of 12 periods (and 12 possible agreements), there are 7 where both Dave and Maureen agreed that agreement did take place (Periods 1, 3, 4, 5, 7, 8, and 12), and 3 where they agreed it did not (Periods 2, 6, and 9), for a total of 10 agreements and 2 disagreements. Interrater reliability is computed using the following simple formula:

$$Interrater\ Reliability = \frac{Number\ of\ Agreements}{Number\ of\ Possible\ Agreements}, \quad (6.6)$$

and when we plug in the numbers as you see here,

$$Interrater\ Reliability = \frac{10}{12} = .833, \quad (6.7)$$

the resulting interrater reliability coefficient is .833.

How Big Is Big? Interpreting Reliability Coefficients

OK, now we get down to business, and guess what? Remember all you learned about interpreting the value of the correlation coefficient in Chapter 5? It's almost the same as when you interpret reliability coefficients as well, with a (little) bit of a difference. We want only two things, and here they are:

- reliability coefficients to be positive (or direct) and not to be negative (or indirect) and
- reliability coefficients that are as large as possible (between .00 and +1.00).

And If You Can't Establish Reliability . . . Then What?

The road to establishing the reliability of a test is not a smooth one at all, and one that takes a good deal of work. What if the test is not reliable?

Here are a few things to keep in mind. Remember that reliability is a function of how much error contributes to the observed score. Lower that error, and you increase the reliability.

- Make sure that the instructions are standardized across all settings when the test is administered.
- Increase the number of items or observations, because the larger the sample from the universe of behaviors you are

investigating, the more likely the sample is representative and reliable. This is especially true for achievement tests.

- Delete unclear items, because some people will respond in one way and others will respond in a different fashion, regardless of their knowledge, ability level, or individual traits.

- For achievement tests especially (such as spelling or history tests), moderate the easiness and difficulty of tests, because any test that is too difficult or too easy does not reflect an accurate picture of one's performance.

- Minimize the effects of external events and standardize directions so that if a particularly important event, such as Mardi Gras or graduation, occurs near the time of testing, you can postpone any assessment.

Just One More Big Thing

The first step in creating an instrument that has sound psychometric (how's that for a big word?) properties is to establish its reliability (and we just spent some good time on that). Why? Well, if a test or measurement instrument is not reliable, is not consistent, and does not do the same thing time after time after time, it does not matter what it measures (and that's the validity question), right?

You could easily have the KACAS (Kids Are Cool at Spelling) test of introductory spelling and the first three items could be

$$16 + 12 = ?$$

$$21 + 13 = ?$$

$$41 + 33 = ?$$

This is surely a highly reliable test, but surely not a valid one. Now that we have reliability well understood, let's move on to an introduction to validity.

VALIDITY—WHOA! WHAT IS THE TRUTH?

Validity is, most simply, the property of an assessment tool that indicates that the tool does what it says it does. A valid test is a test that measures what it is supposed to. If an achievement test is supposed to measure knowledge of history, then that's what it does. If an intelligence test is supposed to measure whatever intelligence is defined as by the test's creators, then it does just that.

Different Types of Validity

Just as there are different types of reliability, so there are different types of validity, and we'll cover the three most important categories and most often used in this section. They are all summarized in Table 6.2.

Table 6.2	Different Types of Validity, When They Are Used, How They Are Computed, and What They Mean		
Type of Validity	*When You Use It*	*How You Do It*	*An Example of What You Can Say When You're Done*
Content validity	When you want to know whether a sample of items truly reflects an entire universe of items in a certain topic	Ask Mr. or Ms. Expert to make a judgment that the test items reflect the universe of items in the topic being measured.	My weekly quiz in my stat class fairly assesses the chapter's content.
Criterion validity	When you want to know if test scores are systematically related to other criteria that indicate the test taker is competent in a certain area	Correlate the scores from the test with some other measure that is already valid and assesses the same set of abilities.	The EATS test (of culinary skills) has been shown to be correlated with being a fine chef 2 years after culinary school (an example of predictive validity).
Construct validity	When you want to know if a test measures some underlying psychological construct	Correlate the set of test scores with some theorized outcome that reflects the construct for which the test is being designed.	It's true—men who participate in body contact and physically dangerous sports score higher on the TEST (osterone) test of aggression.

Content Validity

Content validity is the property of a test such that the test items sample the universe of items for which the test is designed. Content validity is most often used with achievement tests (e.g., everything from your first-grade spelling test to the SATs).

Establishing Content Validity. Establishing content validity is actually very easy. All you need is to locate your local cooperative content expert. For example, if I were designing a test of introductory physics, I would go to the local physics expert (perhaps the teacher at the local high school or a professor at the university who teaches physics) and I would say, "Hey, Albert (or Alberta), do you think this set of 100 multiple-choice items accurately reflects all the possible topics and ideas that I would expect the students in my introductory class to understand?"

I would probably tell Albert or Alberta what the topics were, and then he or she would look at the items and basically provide a judgment as to whether the items meet the criterion I established—a representation of the entire universe of all items that are introductory. If the answer is yes, I'm done (at least for now). If the answer is no, it's back to the drawing board and the creation of new items or the refinement of existing ones.

Criterion Validity

Criterion validity assesses whether a test reflects a set of abilities in a current or future setting. If the criterion is taking place in the here and now, we talk about **concurrent validity.** If the criterion is taking place in the future, we talk about **predictive validity.** For criterion validity to be present, one need not establish both concurrent and predictive validity, only the one that works for the purposes of the test.

Establishing Concurrent Validity. For example, you've been hired by the Universal Culinary Institute to design an instrument that measures culinary skills. Some part of culinary training has to do with straight knowledge. (For example, what's a roux? And that's left to the achievement test side of things.)

So, you develop a test that you think does a good job of measuring culinary skills, and now you want to establish the level of concurrent validity. To do this, you design the COOK scale, a set of 5-point items across a set of criteria (presentation, cleanliness, etc.) that each judge will use. As a criterion (and that's the key here), you have another set of judges rank each student from 1 to 100 on overall ability. Then, you simply correlate the COOK scores with the judge's rankings. If the validity coefficient (a simple correlation) is high, you're in business—if not, it's back to the drawing board.

Establishing Predictive Validity. Let's say that the cooking school has been percolating (heh-heh) along just fine for 10 years and you are interested not only in how well people cook (and that's the concurrent validity part of this exercise that you just established) but in the predictive validity as well. Now, the criterion changes from a here-and-now score (the one that judges give) to one that looks to the future.

Here, we are interested in developing a test that predicts success as a chef 10 years down the line. To establish the predictive validity of the COOK test, you go back and locate graduates of the program who have been out cooking for 10 years and administer the test to them. The criterion that is used here is their level of success, and you use as measures (a) whether they own their own restaurant, and (b) whether it has been in business for more than 1 year (given that the failure rate for new restaurants is more than 80% within the first year). The rationale here is that if a restaurant is in business for more than 1 year, then the chef must be doing something right.

To complete this exercise, you correlate the COOK score with a value of 1 (if the restaurant is in business for more than a year and owned by the graduate) with the previous (10 years earlier) COOK score. A high coefficient indicates predictive validity, and a low correlation indicates the lack thereof.

Construct Validity

Construct validity is the most interesting and the most difficult of all the validities to develop because it is based on some underlying construct or idea behind a test or measurement tool.

You may remember from your extensive studies in Psych 1 that a construct is a group of interrelated variables. For example, aggression is a construct (consisting of such variables as inappropriate touching, violence, lack of successful social interaction, etc.), as is intelligence, mother–infant attachment, and hope. And keep in mind that these constructs are generated from some theoretical position that the researcher assumes. For example, he or she might propose that aggressive men are more often in trouble with the authorities than nonaggressive men.

Establishing Construct Validity. So, you have the FIGHT test (of aggression), which is an observational tool that consists of a series of items that are an outgrowth of your theoretical view about what the construct of aggression consists of. You know from the criminology literature that males who are aggressive do certain types of things more than others—for example, they get into more arguments, they are more physically aggressive (pushing and such), they commit more crimes of violence against others, and they have fewer successful interpersonal relationships. The FIGHT scale includes items that describe different behaviors, some of them theoretically related to aggressive behaviors and some that are not. Once the FIGHT scale is completed, you examine the results to see if positive scores on the FIGHT correlate with the presence of the kinds of behaviors you would predict (level of involvement in crime, quality of personal relationships, etc.) and don't correlate with the kinds of behaviors that should not be related (such as lack of domestic violence, completion of high school and college, etc.). And if the correlation is high for the items that you predict should correlate and low for the items that should not, then you can conclude that there is something about the FIGHT scale (and it is probably the items you designed that do not assess elements of aggression) that works. Congratulations.

AND IF YOU CAN'T ESTABLISH VALIDITY . . . THEN WHAT?

Well, this is a tough one, especially because there are so many different types of validity.

In general, if you don't have the validity evidence you want, your test is not doing what it should. If it's an achievement test, and a satisfactory level of content validity is what you seek, then you probably have to redo the questions on your test to make sure they are more consistent with what they should be according to that expert.

If you are concerned with criterion validity, then you probably need to reexamine the nature of the items on the test and answer the question how well you would expect these responses to these questions to relate to the criterion you selected.

And finally, if it is construct validity that you are seeking and can't seem to find—better take a close look at the theoretical rationale that underlies the test you developed. Perhaps our definition and model of aggression is wrong, or perhaps intelligence needs some critical rethinking.

A Last, Friendly Word

This measurement stuff is pretty cool—intellectually interesting, and, in these times of accountability, everyone wants to know about the progress of students, stockbrokers, social welfare agency programs, and more.

Because of this strong and growing interest, there's a great temptation for undergraduate students working on their honors thesis or semester project or graduate students working on their thesis or dissertation to design an instrument for their final project.

But beware that what sounds like a good idea might lead to a disaster. The process of establishing the reliability and validity of any instrument can take years of intensive work. And what can make matters even worse is when the naïve or unsuspecting individual wants to create a new instrument to test a new hypothesis. That means that on top of everything else that comes with testing a new hypothesis, there is also the task of making sure the instrument works as well.

> If you are doing original research of your own, such as for your thesis or dissertation requirement, be sure to find a measure that has already had reliability and validity evidence well established. That way, you can get on with the main task of testing your hypotheses and not fool with the huge task of instrument development—a career in and of itself. Want a good start? Try the Buros Institute of Mental Measurements, available online at http://www.unl.edu/buros.

VALIDITY AND RELIABILITY: REALLY CLOSE COUSINS

Let's step back for a moment and recall one of the reasons that you're even reading this chapter.

It was assigned to you. No, really. This chapter is important because you need to know something about reliability and the validity of the instruments you are using to measure outcomes. Why? If these instruments are not reliable and valid, then the results of your experiment will always be in doubt.

As we mentioned earlier in this chapter, you can have a test that is reliable, but one that is not valid. However, you cannot have a valid test without it first being reliable. Why? Well, a test can do whatever it does over and over (that's reliability), but still will not do what it is supposed to (that's validity). But, if a test does what it is supposed to, then it has to do it consistently to work.

TECH TALK You've read about the relationship between reliability and validity several places in this chapter, but there's a very cool relationship lurking out there that you may read about later in your coursework that you should know about now. This relationship says that the maximum level of validity is equal to the square root of the reliability coefficient. For example, if the reliability coefficient for a test of mechanical aptitude is .87, the validity coefficient can be no larger than .93 (which is the square root of .87). What this means in tech talk is that the validity of a test is constrained by how reliable it is. And that makes perfect sense if we stop to think that a test must do what it does consistently before we are sure it does what it says it does.

But the relationship is closer as well. You cannot have a valid instrument without it first being reliable, because in order for something to do what it is supposed to do, it must first do it consistently, right? So, the two work hand in hand.

Summary

Yep, this is a stat course, so what's the measurement stuff doing here? Once again, almost any use of statistics revolves around some outcome being measured. Just as you read basic stats to make sense out of lots of data, you need basic measurement information to make sense of how behaviors, test scores, rankings, and whatever else is measured, is assessed.

Time to Practice

1. Go to the library and find five journal articles in your area of interest where reliability and validity data are reported, and discuss the outcome measures that are used. Identify the type of reliability that was established and the type of validity, and comment on whether you think that the levels are acceptable. If not, how can they be improved?

2. Provide an example of when you would want to establish test-retest and parallel forms reliability.

3. You are developing an instrument that measures vocational preferences (what you want to be), and you need to administer the test several times during the year while students are attending a vocational program. You need to assess the test-retest reliability of the test and the data from two administrations (available as Chapter 6 Data Set 1)— one in the fall and one in the spring. Would you call this a reliable test? Why or why not?

4. How can a test be reliable and not valid, and not valid unless it is reliable?

5. When testing any experimental hypothesis, why is it important that the test you use to measure the outcome be both reliable and valid?

Answers to Practice Questions

1. Do this one on your own.

2. First, test–retest. . . . You want to examine the stability of a test score over time and would need to do that if you were testing children's skills in August when school begins, and in May, when it ends. As far as parallel forms, you may want to give two groups of adolescents a test of the same phenomenon, expressed through paper-and-pencil writings and also through the use of word processing software. You want to compare the two and make sure that the evaluating tool is accurate and truthful.

3. The test–retest reliability is less than .14—pretty low to claim any reliability.

4. In order for a test to be valid, it has to do the same thing consistently (which is reliability). If it does not, how can one say it does what it is supposed to do? One can't, my good friend.

5. Ah, the million-dollar question. If one does not use an instrument that is reliable and valid, how does one know that it is the hypothesis not being (or being) supported versus the poor quality of the instrumentation that might yield positive or negative results?

PART III

Taking Chances for Fun and Profit

Snapshots

A scatterplot of student test scores.

What do you know so far, and what's next? To begin with, you've got a really solid basis for understanding how to describe the characteristics of a set of scores and how distributions can differ from one another. That's what you learned in Chapters 2, 3, and 4 of *Statistics for People Who (Think They) Hate Statistics . . . Excel 2007 Edition*. In Chapter 5, you also learned how to describe the relationship between variables using correlational tools, and in Chapter 6, you learned about how to judge the usefulness of assessment tools.

Now it's time to bump up the ante a bit and start playing for real. In Part III of *Statistics for People Who (Think They) Hate Statistics . . . Excel 2007 Edition*, you will be introduced in Chapter 7 to the importance, and nature, of hypothesis testing, including an in-depth discussion of what a hypothesis is, what different types there are, the function of the hypothesis, and why and how hypotheses are tested.

Then we'll get to the all-important topic of probability represented by our discussion of the normal curve and the basic principles underlying probability—the part of statistics that helps us define how likely it is that some event (such as a specific score on a test) will occur. We'll use the normal curve as a basis for these arguments, and you'll see how any score or occurrence within any distribution has a likelihood associated with it.

After some fun with probability and the normal curve, we'll be ready to start our extended discussion in Part IV regarding the application of hypothesis testing and probability theory to the testing of specific questions regarding relationships between variables. It only gets better from here!

7 Hypotheticals and You

Testing Your Questions

Difficulty Scale ☺☺☺ (don't plan on going out tonight)

How much Excel? None

What you'll learn about in this chapter

- The difference between a sample and a population (again)
- The importance of the null and research hypotheses
- The criteria for judging a good hypothesis

SO YOU WANT TO BE A SCIENTIST . . .

You might have heard the term *hypothesis* used in other classes. You may even have had to formulate one for a research project you did for another class, or you may have read one or two in a journal article. If so, then you probably have a good idea what a hypothesis is. For those of you who are unfamiliar with this often-used term, a **hypothesis** is basically "an educated guess." Its most important role is to reflect the general problem statement or question that was the motivation for asking the research question in the first place.

More Excel

Chapter 7 is one of the few times in *Statistics for People* . . . when you will not be using Excel to actively learn about basic statistics. This chapter is pretty much full of ideas that are important to understand, and we will use Excel again in later chapters (beginning with the next one) to illustrate some of those ideas.

That's why taking the care and time to formulate a really precise and clear research question is so important. This research question will be your guide in the creation of a hypothesis, and in turn, the hypothesis will determine the techniques you will use to test the hypothesis and answer the question that was originally asked.

So, a good hypothesis translates a problem statement or a research question into a form that is more amenable to testing. This form is called a hypothesis. We talk about what makes a good hypothesis later in this chapter. Before that, let's turn our attention to the difference between a sample and a population. This is an important distinction because hypothesis testing deals with a sample, and then the results are generalized to the larger population. We also address the two main categories of hypotheses (the null hypothesis and the research hypothesis). But first, let's formally define some simple terms that we have used earlier in *Statistics for People Who (Think They) Hate Statistics . . . Excel 2007 Edition.*

Samples and Populations

As a good scientist, you would like to be able to say that if Method A is better than Method B, this is true forever and always and for all people in the universe, right? Indeed. And, if you do enough research on the relative merits of Methods A and B and test enough people, you may someday be able to say that. But don't get too excited, because it's unlikely that you will be able to speak with such confidence. It takes too much money ($$$) and too much time (all those people!) to do all that research, and besides, it's not even necessary. Instead, you can just select a representative sample from the population and test your hypothesis about Methods A and B.

Given the constraints of never enough time and never enough research funds, with which almost all scientists live, the next best strategy is to take a portion of a larger group of participants and do the research with that smaller group. In this context, the larger group is referred to as a **population,** and the smaller group selected from that population is referred to as a **sample.**

 TECH TALK A measure of how well a sample approximates the characteristics of a population is called **sampling error.** Sampling error is basically the difference between the values of the sample statistic and the population parameter. The higher the sampling error, the less precision one has in sampling and the more difficult it will be to make the case that what you find in the sample indeed reflects what you expect to find in the population.

Samples should be selected from populations in such a way that the sample matches as closely as possible the characteristics of the population. The goal is to have the sample as much like the population as possible. The most important implication of ensuring similarity between the two is that the research results based on the sample can be generalized to the population. When the sample accurately represents the population, the results of the study are said to have a high degree of generalizability.

A high degree of generalizability is an important quality of good research because it means that the time and effort (and $$$) that went into the research may have implications for groups of people other than the original participants.

It's easy to equate "big" with "representative." Keep in mind that it is far more important to have a representative sample than it is to have a big sample (people always think that big is better—only true on Thanksgiving). For example, lots and lots of participants in a sample is very impressive, but if the participants do not represent the larger population, the research will have much less value.

THE NULL HYPOTHESIS

OK. So we have a sample of participants selected from a population, and to begin the test of our research hypothesis, we first formulate the **null hypothesis.**

The null hypothesis is an interesting little creature. If it could talk, it would say something like, "I represent no relationship between the variables that you are studying." In other words, null hypotheses are statements of equality demonstrated by the following real-life (brief) null hypotheses taken from a variety of popular social and behavioral science journals. Names have been changed to protect the innocent.

- There will be *no difference* in the average score of 9th graders and the average score of 12th graders on the ABC memory test.
- There is *no difference* between the effectiveness of community-based, long-term care for older adults and the effectiveness of in-home, long-term care on the social activities of older adults.
- There is *no relationship* between reaction time and problem-solving ability.
- There is *no difference* between white and black families in the amount of assistance offered to their children in school-related activities.

What these four null hypotheses have in common is that they all contain a statement that two or more things are equal, or unrelated, to each other.

The Purposes of the Null Hypothesis

What are the basic purposes of the null hypothesis? The null hypothesis acts as both a starting point and a benchmark against which the actual outcomes of a study can be measured. Let's examine each of these purposes in more detail.

First, the null hypothesis acts as a starting point because it is the state of affairs that is accepted as true in the absence of any other information. For example, let's look at the first null hypothesis we stated above:

There will be no difference in the average score of 9th graders and the average score of 12th graders on the ABC memory test.

Given absolutely no other knowledge of 9th and 12th graders' memory skills, you have no reason to believe that there will be differences between the two groups, right? If you know nothing about the relationship between these variables, the best you can do is guess. And that's taking a chance. You might speculate as to why one group might outperform another, but if you have no evidence a priori (before the fact), then what choice do you have but to assume that they are equal?

This lack of a relationship as a starting point is a hallmark of this whole topic. In other words, until you prove that there is a difference, you have to assume that there is no difference. And a statement of no difference or no relationship is exactly what the null hypothesis is all about.

Furthermore, if there are any differences between these two groups, you have to assume that these differences are due to the most attractive explanation for differences between any groups on any variable—chance! That's right; given no other information, chance is always the most likely and attractive explanation for the observed differences between two groups or the relationship between variables. Chance explains what we cannot. You might have thought of chance as the odds on winning that $5,000 nickel jackpot at the slots, but we're talking about chance as all that other "stuff" that clouds the picture and makes it even more difficult to understand the "true" nature of relationships between variables.

For example, you could take a group of soccer players and a group of football players and compare their running speeds. But look at all the factors we don't know about that could contribute to differences. Who is to know whether some soccer players practice more, or if some football players are stronger, or if both groups are receiving additional training? What's more, perhaps the way their speed is being measured leaves room for chance; a faulty stopwatch or a windy day can contribute to differences unrelated to true running speed. As good researchers, our job is to eliminate chance factors from explaining observed differences and to evaluate other factors that might contribute to group differences, such as intentional training or nutrition programs, and see how they affect speed. The point is, if we find differences between groups and the differences are not due to training, we are at a loss as to what to attribute the difference to other than chance.

The second purpose of the null hypothesis is to provide a benchmark against which observed outcomes can be compared to see if these differences are due to some other factor. The null hypothesis helps to define a range within which any observed differences between groups can be attributed to chance (which is the null hypothesis's contention and reason for being born) or are due to something other than chance (which perhaps would be the result of the manipulation of some variable, such as training in the above example).

Most research studies have an implied null hypothesis, and you may not find it clearly stated in a research report or journal article. Instead, you'll find the research hypothesis clearly stated, which is now where we turn our attention.

THE RESEARCH HYPOTHESIS

Whereas a null hypothesis is a statement of no relationship between variables, a **research hypothesis** is a definite statement that there is a relationship between variables. For example, for each of the null hypotheses stated earlier, here is a corresponding research hypothesis. Notice that we said "a" and not "the" corresponding research hypothesis, because there certainly could be more than one research hypothesis for any one null hypothesis.

- The average score of 9th graders *is different* from the average score of 12th graders on the ABC memory test.
- The effectiveness of community-based, long-term care for older adults *is different* from the effectiveness of in-home, long-term care on the social activities of older adults when measured using the Margolis Scale of Social Activities.
- Slower reaction time and problem-solving ability *are positively related.*

- There *is a difference* between white and black families in the amount of assistance offered to their children in educational activities.

Each of these four research hypotheses has one thing in common. They are all statements of *inequality*. They posit a relationship between variables and not equality, as does the null hypothesis.

The nature of this inequality can take two different forms—a directional or a nondirectional research hypothesis. If the research hypothesis posits no direction to the inequality (such as "different from"), the hypothesis is a nondirectional research hypothesis. If the research hypothesis posits a direction to the inequality (such as "more than" or "less than"), the research hypothesis is a directional research hypothesis.

The Nondirectional Research Hypothesis

A **nondirectional research hypothesis** reflects a difference between groups, but the direction of the difference is not specified. For example, the research hypothesis

The average score of 9th graders is different from the average score of 12th graders on the ABC memory test

is nondirectional in that the direction of the difference between the two groups is not specified. The hypothesis states only that there is a difference and says nothing about the direction of that difference. It is a research hypothesis because a difference is hypothesized, but the nature (in other words, the direction) of the difference is not specified.

A nondirectional research hypothesis such as the one described here would be represented by the following equation:

$$H_1 : \bar{X}_9 \neq \bar{X}_{12} , \qquad (7.1)$$

where

H_1 represents the symbol for the first (of possibly several) research hypotheses,

$\bar{X}_9$ represents the average memory score for the sample of 9th graders,

$\bar{X}_{12}$ represents the average memory score for the sample of 12th graders, and

$\neq$ means "is not equal to".

The Directional Research Hypothesis

A **directional research hypothesis** reflects a difference between groups, and the direction of the difference is specified.

For example, the research hypothesis

The average score of 12th graders is greater than the average score of 9th graders on the ABC memory test

is directional because the direction of the difference between the two groups is specified. One is hypothesized to be greater than (not just different from) the other.

An example of two other directional hypotheses is

A is greater than B (or A > B)

and

B is greater than A (or A < B).

These both represent inequalities, but of a specific nature (greater than or less than). A directional research hypothesis such as the one described above, where 12th graders are hypothesized to score better than 9th graders, would be represented by the following equation:

$$H_1 : \bar{X}_{12} > \bar{X}_9, \tag{7.2}$$

where

H_1 represents the symbol for the first (of possibly several) research hypotheses,

$\bar{X}_9$ represents the average memory score for the sample of 9th graders,

$\bar{X}_{12}$ represents the average memory score for the sample of 12th graders, and

>   means "is greater than."

What is the purpose of the research hypothesis? It is this hypothesis that is directly tested as an important step in the research process. The results of this test are compared with what you expect by chance alone (reflecting the null hypothesis) to see which of the two is the more attractive explanation for any differences between groups you might observe.

Here are the four null hypotheses stated as both directional and nondirectional research hypotheses.

Table 7.1 Null Hypotheses and Corresponding Research Hypotheses

Null Hypothesis	Nondirectional Research Hypothesis	Directional Research Hypothesis
There will be no difference in the average score of 9th graders and the average score of 12th graders on the ABC memory test.	Twelfth graders and 9th graders will differ on the ABC memory test.	Twelfth graders will have a higher average score on the ABC memory test than will 9th graders.
There is no difference between the effectiveness of community-based, long-term care for older adults and the effectiveness of in-home, long-term care on the Margolis Scale of Social Activities in older adults.	The effect of community-based, long-term care for older adults is different from the effect of in-home, long-term care on the social activities of older adults when measured using the Margolis Scale of Social Activities.	Older adults exposed to community-based, long-term care score higher on the Margolis Scale of Social Activities than do older adults receiving in-home, long-term care.
There is no relationship between reaction time and problem-solving ability.	There is a relationship between reaction time and problem-solving ability.	There is a positive relationship between reaction time and problem-solving ability.
There is no difference between white and black families in the amount of assistance offered to their children.	The amount of assistance offered by white families to their children is different from the amount of support offered by black families to their children.	The amount of assistance offered by white families to their children is more than the amount of support offered by black families to their children.

TECH TALK

What About Those Tails?

Another way to talk about directional and nondirectional hypotheses is to talk about one- and two-tailed tests. A **one-tailed test** (reflecting a directional hypothesis) posits a difference in a particular direction, such as when we hypothesize that Group 1 will score higher than Group 2. A **two-tailed test** (reflecting a nondirectional hypothesis) posits a difference but in no particular direction. The importance of this distinction begins when you test different types of hypotheses (one- and two-tailed) and establish probability levels for rejecting or not rejecting the null hypothesis. More about this in Chapter 10. Promise.

Some Differences Between the Null Hypothesis and the Research Hypothesis

Besides the null hypothesis representing an equality and the research hypothesis representing an inequality, there are several other important differences between the two types of hypotheses.

First, for a bit of review, the two types of hypotheses differ in that one (the null hypothesis) states that there is no relationship between variables (an equality), whereas the research hypothesis states that there is a relationship between the variables (an inequality). This is the primary difference.

Second, null hypotheses always refer to the population, whereas research hypotheses always refer to the sample. We select a sample of participants from a much larger population. We then try to generalize the results from the sample back to the population. If you remember your basic philosophy and logic (you did take these courses, right?), you'll remember that going from small (as in a sample) to large (as in a population) is a process of inference.

Third, because the entire population cannot be tested directly (again, it is impractical, uneconomical, and often impossible), you can't say with 100% certainty that there is no real difference between samples on some variable. Rather, you have to infer it (indirectly) from the results of the test of the research hypothesis, which is based on the sample. Hence, the null hypothesis is, by definition, indirectly tested, and the research hypothesis is directly tested. Fourth, null hypotheses are always written using Greek symbols, and research hypotheses are always written using Roman symbols. For example, the null hypothesis that the average score for 9th graders is equal to that of 12th graders is represented as you see here:

$$H_0 : \mu_9 = \mu_{12}, \tag{7.3}$$

where

H_0 represents the null hypothesis,

μ_9 represents the theoretical average for the population of 9th graders, and

μ_{12} represents the theoretical average for the population of 12th graders.

The research hypothesis that the average score for a sample of 12th graders is greater than the average score for a sample of 9th graders is shown in Formula 7.2.

Finally, because you cannot directly test the null hypothesis, it is an implied hypothesis. But the research hypothesis is explicit and is stated as such. This is another reason why you rarely see null hypotheses stated in research reports and almost always see a statement (be it in symbols or words) of the research hypothesis.

WHAT MAKES A GOOD HYPOTHESIS?

You now know that hypotheses are educated guesses—a starting point for a lot more to come. As with any guess, some are better than others right from the start. We can't stress enough how important it is to ask the question you want answered and to keep in mind that any hypothesis you present is a direct extension of the original question you asked. This question will reflect your own personal interests and motivation and the research that has been done previously. With that in mind, here are criteria you might use to decide whether a hypothesis you read in a research report or the ones you formulate are acceptable.

To illustrate, let's use an example of a study that examines the effects of after-school child care for employees who work late on the parents' adjustment to work. Here is a well-written hypothesis:

Parents who enroll their children in after-school programs will miss fewer days of work in one year and will have a more positive attitude toward work, as measured by the Attitude Toward Work survey, than will parents who do not enroll their children in such programs.

Here are the criteria.

First, a good hypothesis is stated in declarative form and not as a question. In the above example, the question, "Do you think parents and the companies they work for will be better . . . ?" was not posed because hypotheses are most effective when they make a clear and forceful statement.

Second, a good hypothesis posits an expected relationship between variables. The hypothesis that is being used as an example clearly describes the relationship between after-school child care, parents' attitude, and absentee rate. These variables are being tested to see if one (enrollment in the after-school program) has an effect upon the others (absentee rate and attitude).

Notice the word "expected" in the above criterion? Defining an expected relationship is intended to prevent the fishing trip (sometimes called the "shotgun" approach) that may be tempting to take but is not very productive. You do get somewhere using the shotgun approach, but because you don't know where you started, you have no idea where you end up.

TECH TALK The fishing trip approach is where you throw out your line and take anything that bites. You collect data on as many things as you can, regardless of your interest or even whether collecting the data is a reasonable part of a scientific investigation. Or, you load up them guns and blast away at anything that moves and you're bound to hit something. The problem is, you may not want what you hit, and, worse, you may miss what you want to hit, and worst of all (if possible), you may not know what you hit! Good researchers do not want just anything they can catch or shoot. They want specific results. To get them, researchers need their opening questions and hypotheses to be clear, forceful, and easily understood.

Third, hypotheses reflect the theory or literature on which they are based. As you read in Chapter 1, the accomplishments of scientists rarely can be attributed to just their own hard work. Their accomplishments are always due, in part, to many other researchers who came before them and laid the framework for later explorations. A good hypothesis reflects this, in that it has a substantive link to existing literature and theory. In the above example, let's assume there is literature indicating that parents are more comfortable knowing their children are being cared for in a structured environment, and parents can then be more productive at work. Knowing this would allow one to hypothesize that an after-school program would provide the security for which parents are looking. In turn, this allows them to concentrate on working rather than calling on the telephone to find out whether Rachel or Gregory got home safely.

Fourth, a hypothesis should be brief and to the point. You want your hypothesis to describe the relationship between variables in a declarative form and to be as direct and explicit as possible. The more to the point it is, the easier it will be for others (such as your master's thesis or doctoral dissertation committee members!) to read your research and understand exactly what you are hypothesizing and what the important variables are. In fact, when people read and evaluate research (as you will learn more about later in this chapter), the first thing many of them do is find the hypotheses to get a good idea as to the general purpose of the research and how things will be done. A good hypothesis tells you both of these things.

Fifth, good hypotheses are testable hypotheses—and testable hypotheses contain variables that can be measured. This means that you can actually carry out the intent of the question reflected by the hypothesis. You can see from the sample hypothesis that the important comparison is between parents who have enrolled their child in an after-school program and those who have not. Then, such things as attitude and work days missed will be measured. These are both reasonable objectives. Attitude is measured by the Attitude Toward Work survey (a fictitious title, but you get the idea), and absenteeism (the number of days away from work) is an easily recorded and unambiguous measure. Think how much harder things would be if the hypothesis were stated as *Parents who enroll their children in after-school care feel better about their jobs.* Although you might get the same message, the results might be more difficult to interpret given the ambiguous nature of words such as "feel better."

In sum, hypotheses should

- be stated in declarative form,
- posit a relationship between variables,
- reflect a theory or a body of literature on which they are based,
- be brief and to the point, and
- be testable.

When a hypothesis meets each of these five criteria, you know that it is good enough to continue with a study that will accurately test the general question from which the hypothesis was derived.

Summary

A central component of any scientific study is the hypothesis, and the different types of hypotheses (null and research) help form a plan for answering the questions asked by the purpose of our research. The starting point and benchmark that characterize the null hypothesis let us use it as a comparison as we evaluate the acceptability of the research hypothesis. Now let's move on to how those null hypotheses are actually tested.

Time to Practice

1. Go to the library and select five empirical (those containing data) research articles from your area of interest. For each one, list the following:
   a. What is the null hypothesis (implied or explicitly stated)?
   b. What is the research hypothesis (implied or explicitly stated)?
   c. In your own area of interest, create a null and research hypothesis.
   d. And what about those articles with no hypothesis clearly stated or implied? Identify those articles and see if you can write a research hypothesis.

2. For the following research questions, create one null hypothesis, one directional research hypothesis, and one nondirectional research hypothesis.
    a. What are the effects of attention on out-of-seat classroom behavior?
    b. What is the relationship between the quality of a marriage and the quality of the spouses' relationships with their siblings?
    c. What's the best way to treat an eating disorder?

3. Go back to the five hypotheses that you found in Question 1 above, and evaluate each using the five criteria that were discussed at the end of the chapter.

4. Why does the null hypothesis presume no relationship between variables?

Answers to Practice Questions

Questions 1 and 3 are specific to your own interests. So, although there are no right answers, there are plenty of wrong ones!

2a. *Null:* Children with short attention spans, as measured by the Attention Span Observation Scale, will have the same frequency of out-of-seat behavior as those with long attention spans.

Directional: Children with short attention spans, as measured by the Attention Span Observation Scale, will have a higher frequency of out-of-seat behavior than those with long attention spans.

Nondirectional: Children with short attention spans, as measured by the Attention Span Observation Scale, differ in the frequency of out-of-seat behavior from those with long attention spans.

2b. *Null:* There is no relationship between the overall quality of a marriage and spouses' relationships with their siblings.

Directional: There is a positive relationship between the overall quality of a marriage and spouses' relationships with their siblings.

Nondirectional: There is a relationship between the overall quality of a marriage and spouses' relationships with their siblings.

2c. *Null:* Pharmacological treatment combined with traditional psychotherapy has the same effect in treating anorexia nervosa as does traditional psychotherapy alone.

Directional: Pharmacological treatment combined with traditional psychotherapy is more effective in treating anorexia nervosa than is traditional psychotherapy alone.

Nondirectional: Pharmacological treatment combined with traditional psychotherapy is different from treating anorexia nervosa with traditional psychotherapy alone.

4. If you're at the beginning of exploring a question (which then becomes a hypothesis) and you have little knowledge about the outcome (which is why you are asking the question and performing the test), then the null is the perfect starting point because it is a statement of equality that basically says, "Given no other information about the relationships that we are studying, I should start at the beginning where I know very little." The null is the perfect, unbiased, and objective starting point because it is the place where everything is thought to be equal unless proved otherwise.

8

Are Your Curves Normal?

Probability and Why It Counts

Difficulty Scale ☺☺☺☺ (not too easy and not too hard, but very important)

How much Excel? 📊 📊 📊 (lots)

What you'll learn about in this chapter

- Why understanding probability is basic to understanding statistics
- What the normal, or bell-shaped, curve is and what its characteristics are
- How to compute and interpret z scores

WHY PROBABILITY?

And here you thought this was a statistics class! Ha! Well, as you will learn in this chapter, the study of probability is the basis for the normal curve (much more on that later) and the foundation for inferential statistics.

Why? First, the normal curve provides us with a basis for understanding the probability associated with any possible outcome (such as the odds of getting a certain score on a test or the odds of getting a head on one flip of a coin).

Second, the study of probability is the basis for determining the degree of confidence we have in stating that a particular finding or outcome is "true." Or, better said, that an outcome (such as an average score) may not have occurred because of chance alone. For example, let's compare Group A (which participates in 3 hours of extra swim practice each week) and Group B (which has no extra swim practice each week). We find that Group A differs from Group B on a test of fitness, but can we say that the difference is due to the extra practice or due to something else? The tools that the study of probability provides allow us to determine the exact mathematical likelihood that the difference is due to practice versus something else (such as chance).

All that time we spent on hypotheses in the previous chapter was time well spent. Once we put together our understanding of what a null hypothesis and a research hypothesis are with the ideas that are the foundation of probability, we'll be in a position to discuss how likely certain outcomes (formulated by the research hypothesis) are.

THE NORMAL CURVE (A.K.A. THE BELL-SHAPED CURVE)

What is a normal curve? Well, the **normal curve** (also called a **bell-shaped curve,** or bell curve) is a visual representation of a distribution of scores that has three characteristics. Each of these characteristics is illustrated in Figure 8.1.

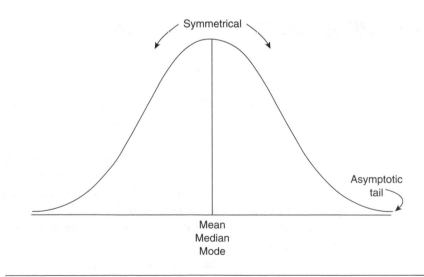

Figure 8.1   The Normal, or Bell-Shaped, Curve

The normal curve represents a distribution of values where *the mean, median, and mode are equal to one another.* You probably remember from Chapter 4 that if the median and the mean are different, then the distribution is skewed in one direction or the other. The normal curve is not skewed. It's got a nice hump (only one), and that hump is right in the middle.

Second, the normal curve is *perfectly symmetrical about the mean.* If you fold one half of the curve along its center line, the two halves would fit perfectly on each other. They are identical. One half of the curve is a mirror image of the other.

Finally (and get ready for a mouthful), *the tails of the normal curve are* **asymptotic**—a big word. What it means is that they come closer and closer to the horizontal axis, but never touch.

The normal curve's shape of a bell also gives the graph its other name, the bell-shaped curve.

TECH TALK

When your devoted author was knee-high, he always wondered how the tail of a normal curve can approach the horizontal or *x*-axis yet never touch it. Try this. Place two pencils one inch apart and then move them closer (by half) so they are one-half inch apart, and then closer (one-quarter inch apart), and closer (one-eighth inch apart). They continually get closer, right? But they never (and never will) touch. Same thing with the tails of the curve. The tail slowly approaches the axis on which the curve "rests," but they can never really touch.

Why is this important? As you will learn later in this chapter, the fact that the tails never touch means that there is an infinitely small likelihood that a score can be obtained that is very extreme (way out in the left or right tail of the curve). If the tails did touch, then the likelihood that a very extreme score could be obtained would be nonexistent.

Hey, That's Not Normal!

We hope your next question is, "But there are plenty of sets of scores where the distribution is not normal or bell shaped, right?" Yes (and here comes the big *but*). When we deal with large sets of data (more than 30), and as we take repeated samples of the data from a population, the values in the curve closely approximate the shape of a normal curve. This is very important, because much of what we do when we talk about inferring from a sample to a population is based on the assumption that what is taken from a population is distributed normally.

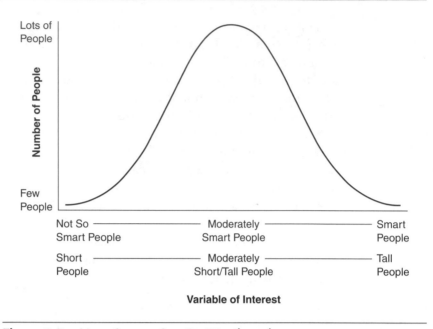

Figure 8.2   How Scores Can Be Distributed

And as it turns out, in nature in general, many things are distributed with the characteristics of what we call normal. That is, there are lots of events or occurrences right in the middle of the distribution, but relatively few on each end, as you can see in Figure 8.2, which shows the distribution of IQ and height in the general population.

For example, there are very few people who are "smart people" (see the right end of the curve) and very few who are "not-so-smart people" (see the left end of the curve). There are lots who are right in the middle ("moderately smart people") and fewer as we move toward the tails of the curve.

Another example is that there are relatively few tall people and relatively few short people, but lots of people of moderate height right in the middle. In both of these examples, the distribution of intellectual skills and height approximate a normal distribution. Consequently, those events that tend to occur in the extremes of the normal curve have a smaller probability associated with each occurrence. We can say with a great deal of confidence that the odds of any one person (whose height we do not know beforehand) being very tall (or very short) are just not very great. But we know that the odds of any one person being average in height, or right around the middle, are pretty good (yikes—that's why it's called the average!). Those events that tend to occur in the middle of the normal curve have a higher probability of occurring than do those in the extremes.

More Normal Curve 101

You already know the three main characteristics that make a curve normal or make it appear bell shaped, but there's more to it than that. Take a look at the curve in Figure 8.3.

The distribution represented here has a mean of 100 and a standard deviation of 10. We've added numbers across the *x*-axis that represent the distance in standard deviations from the mean for this distribution. You can see that the *x*-axis (representing the scores in the distribution) is marked from 70 through 130 in increments of 10 (which is the standard deviation for the distribution), the value of 1 standard deviation. We made up these numbers (100 and 10), so don't go nuts trying to find out where we got them from.

So, a quick review tells us that this distribution has a mean of 100 and a standard deviation of 10. Each vertical line within the curve separates the curve into a section, and each section is bound by particular scores. For example, the first section to the right of the mean of 100 is bound by the scores 100 and 110 representing 1 standard deviation from the mean (which is 100).

And below each raw score (70, 80, 90, 100, 110, 120, and 130), you'll find a corresponding standard deviation (−3, −2, −1, 0, +1, +2, and +3). As you may have figured out already, each standard deviation in our example is 10 points. So 1 standard deviation from the mean (which is 100) is the mean plus 10 points or 110. Not so hard, is it?

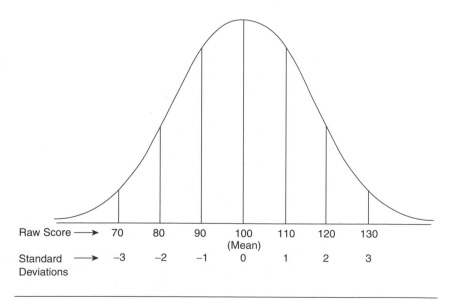

Raw Score ⟶ 70 80 90 100 110 120 130
(Mean)

Standard ⟶ −3 −2 −1 0 1 2 3
Deviations

Figure 8.3 A Normal Curve Divided Into Different Sections

If we extend this argument further, then you should be able to see how the range of scores represented by a normal distribution with a mean of 100 and a standard deviation of 10 is 70 through 130 (which is −3 to +3 standard deviations).

Now, here's a big fact that is always true about normal distributions, means, and standard deviations: For any distribution of scores (regardless of the value of the mean and standard deviation), if the scores are distributed normally, almost 100% of the scores will fit between −3 and +3 standard deviations from the mean. This is very important, because it applies to all normal distributions. Because the rule does apply (once again, regardless of the value of the mean or standard deviation), distributions can be compared with one another. We'll get to that again later.

With all that said, we'll extend our argument a bit more. If the distribution of scores is normal, we can also say that between different points along the x-axis (such as between the mean and 1 standard deviation), a certain percentage of cases will fall. In fact, between the mean (which in this case is 100—got that yet?) and 1 standard deviation above the mean (which is 110), about 34% (actually 34.13%) of all cases in the distribution of scores will fall. This is a fact you can take to the bank because it will always be true.

Want to go further? Take a look at Figure 8.4. Here, you can see the same normal curve in all its glory (the mean equals 100 and the standard deviation equals 10), and the percentage of cases that we would expect to fall within the boundaries defined by the mean and standard deviation.

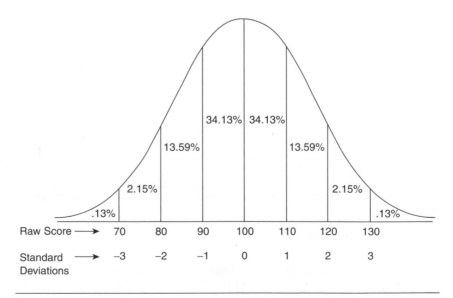

Figure 8.4   Distribution of Cases Under the Normal Curve

Here's what we can conclude.

The distance between	contains	and the scores that are included (if the mean = 100 and the standard deviation = 10) are from
The mean and 1 standard deviation	34.13% of all the cases under the curve	100 to 110
1 and 2 standard deviations	13.59% of all the cases under the curve	110 to 120
2 and 3 standard deviations	2.15% of all the cases under the curve	120 to 130
3 standard deviations and above	0.13% of all the cases under the curve	Above 130

If you add up all the values in either half of the normal curve, guess what you get? That's right, 50%. Why? The distance between the mean and all the scores to the right of the mean underneath the normal curve contains 50% of all the scores.

And because the curve is symmetrical about its central axis (each half is a mirror image of the other), the two halves together represent 100% of all the scores. Not rocket science, but important to point out, nonetheless.

Now, let's extend the same logic to the scores to the left of the mean of 100.

The distance between	contains	and the scores that are included (if the mean = 100 and the standard deviation = 10) are from
The mean and −1 standard deviation	34.13% of all the cases under the curve	90 to 100
−1 and −2 standard deviations	13.59% of all the cases under the curve	80 to 90
−2 and −3 standard deviations	2.15% of all the cases under the curve	70 to 80
−3 standard deviations and below	0.13% of all the cases under the curve	Below 70

Now, be sure to keep in mind that we are using a mean of 100 and a standard deviation of 10 only as sample figures for a particular example. Obviously, not all distributions have a mean of 100 and a standard deviation of 10.

All of this is pretty neat, especially when you consider that the values of 34.13% and 13.59% and so on are absolutely independent of the actual values of the mean and the standard deviation. This value is roughly 34% because of the shape of the curve, not because of the value of any of the scores in the distribution or the value of the mean or standard deviation. In fact, if you actually drew a normal curve on a piece of cardboard and then cut out the area between the mean and +1 standard deviation and then weighed it, it would tip the scale at exactly 34.13% of the weight of the entire piece of cardboard from which the curve was cut. (Try it—it's true.)

In our example, this means that (roughly) 68% (34.13% doubled) of the scores fall between the raw score values of 90 and 110. What about the other 32%? Good question. One half (16%, or 13.59% + 2.15% + 0.13%) falls above (to the right of) 1 standard deviation above the mean and one half falls below (to the left of) 1 standard deviation below the mean. And because the curve slopes, and the amount of area decreases as you move farther away from the mean, it is no surprise that the likelihood that a score will fall more toward the extremes of the distribution is less than the likelihood it will fall toward the middle. That's why the curve has a bump in the middle and is not skewed in either direction.

OUR FAVORITE STANDARD SCORE: THE Z SCORE

You have read more than once how distributions differ in measures of their central tendency and variability.

But in the general practice of research, we will find ourselves working with distributions that are indeed different, yet we will be required to compare them with one another. And to do such a comparison, we need some kind of a standard.

Say hello to **standard scores.** These are scores that are comparable because they are standardized in units of standard deviations. For example, a standard score of 1 in a distribution with a mean of 50 and a standard deviation of 10 means the same as a standard score of 1 from a distribution with a mean of 100 and a standard deviation of 5; they both represent 1 standard score and are an equivalent distance from their respective means. Also, we can use our knowledge of the normal curve and assign a probability to the occurrence of a value that is 1 standard deviation from the mean. We'll do that later.

Although there are other types of standard scores (such as *T* scores), the one that you will see most frequently in your study of statistics is called a **z score.** This is the result of dividing the amount that a raw score differs from the mean of the distribution by the standard deviation (see Formula 8.1).

$$z = \frac{(X - \bar{X})}{s},$$ (8.1)

where

z is the z score,

X is the individual score,

$\bar{X}$ is the mean of the distribution, and

s is the distribution standard deviation.

For example, in Formula 8.2, you can see how the z score is calculated if the mean is 100, the raw score is 110, and the standard deviation is 1.0:

$$z = \frac{(110 - 100)}{10} = +1.0.$$ (8.2)

It's just as easy to compute a raw score given a z score as the other way around. You already know the formula for a z score given the raw score, mean, and standard deviation. But if you know only the z score and the mean and standard deviation, then what's the corresponding raw score? Easy, just use the formula $X = z(s) + \bar{X}$. You can easily convert raw scores to z scores and back again if necessary. For example, a z score of −.5 in a distribution with a mean of 50 and an s of 5 would equal a raw score of $X = (−.5)(5) + 50$, or 47.5.

The following data show the original raw scores plus the z scores for a sample of 10 scores that has a mean of 12 and a standard deviation of 2. Any raw score above the mean will have a corresponding z score that is positive, and any raw score below the mean will have a corresponding z score that is negative. For example, a raw score of 15 has a corresponding z score of +1.5, and a raw score of 8 has a corresponding z score of −2. And of course, a raw score of 12 (or the mean) has a z score of 0 (which it must be because it is no distance from the mean).

X	$X - \bar{X}$	z Score
12	0	0
15	3	1.5
11	−1	−0.5
13	1	0.5
8	−4	−2
14	2	1
12	0	0
13	1	0.5
12	0	0
10	−2	−1

Below are just a few observations about these scores, as a little review.

First, those scores below the mean (such as 8 and 10) have negative z scores, and those scores above the mean (such as 13 and 14) have positive z scores.

Second, positive z scores always fall to the right of the mean and are in the upper half of the distribution. And negative z scores always fall to the left of the mean and are in the lower half of the distribution.

Third, when we talk about a score being located 1 standard deviation above the mean, it's the same as saying that the score is 1 z score above the mean. For our purposes, when comparing scores across distributions, z scores and standard deviations are equivalent. In other words, a z score is simply the number of standard deviations from the mean.

Finally (and this is very important), z scores across different distributions are comparable. Here's another table, similar to the one above, that illustrates this point. These 10 scores were selected from a set of 100 scores, with the scores having a mean of 58 and a standard deviation of 15.3.

Raw Score	$X - \bar{X}$	z Score
67	9	0.59
54	−4	−0.26
68	10	0.65
33	−25	−1.63
56	−2	−0.13
76	18	1.17
65	7	0.46
35	−23	−1.50
48	−8	−0.52
76	18	1.17

In the first distribution you saw earlier, with a mean of 12 and a standard deviation of 2, a raw score of 12.8 has a corresponding z score of +0.4, which means that a raw score of 12.8 is a distance of 0.4 standard deviations from the mean. In the second distribution, with a mean of 58 and a standard deviation of 15.3, a raw score of 64.2 has a corresponding z score of +0.4 as well. A miracle? No—just a good idea.

Both raw scores of 12.8 and 64.2, *relative to one another*, are equal distances from the mean. When these raw scores are represented as standard scores, then they are directly comparable to one another in terms of their relative location in their respective distributions. That is a very cool way to compare outcomes and one that helps level the playing field in comparing performances from one group to another.

Using Excel to Compute z Scores

Using Excel to compute z scores (or any standard score for that matter) is a cinch and requires only the use of the same simple formula. In Figure 8.5, you can see a worksheet of the same data you see on page 184, and the formula that is used to compute the z scores. That formula was then copied from Cell B2 through to Cell B11. Very easy and very quick.

	B2			f_x	=(A2-(AVERAGE($A$2:$A$11)))/STDEV($A$2:$A$11)			
	A	B	C	D	E	F	G	H
1	X	z score						
2	12	0						
3	15	1.5						
4	11	-0.5						
5	13	0.5						
6	8	-2						
7	14	1						
8	12	0						
9	13	0.5						
10	12	0						
11	10	-1						

Figure 8.5   Using Excel to Compute z Scores

We did a few new and neat things here.

First, we used functions within a formula. Instead of using functions to compute the mean and the standard deviation and then using a formula to compute the z scores, we skipped right to using the functions. So, instead of the formula for a z score looking like this,

$$z = \frac{(X - \bar{X})}{s},$$

(8.3)

ours looks like this,

=(A2-(AVERAGE($A$2:$A$11)))/STDEV($A$2:$A$11).

Here, deviation from the mean was computed by taking the raw score (in Cell A2) and subtracting from it the average (computed using the AVERAGE function), and that outcome was then divided by the standard deviation, which we computed using the STDEV function. The advantage of using the second formula is that it saved us the steps of creating separate values that then had to be plugged into a formula.

Second, what's with all those dollar signs $$$$? Simple—remember way back in the introduction we mentioned relative versus absolute references? We entered the cell addresses as absolute references because the cells that are used to compute the average and the standard deviation do not change, whereas the cell in which the computation of the z score occurs does. That's why, in Figure 8.5, you see the cell reference as A2 without dollar signs because it is a relative one and one in which the value will change as the formula is copied down the column.

More Excel

Excel has a cute function (**STANDARDIZE** function) that computes the standard score for any raw score (or, as we show you, any set of raw scores). The function takes the form of

=STANDARDIZE(X,mean,standard deviation),

where

X = the value for which you want to compute a z score,

mean = the mean of the set of scores, and

standard deviation = (guess what) the standard deviation of the set of scores.

For example, if you wanted to compute the z score for a raw score of 98, which is one data point from a set of data points with a mean of 95 and a standard deviation of 2.3 (whew), the function would look like this:

=STANDARDIZE(98,95,2.3).

Figure 8.6 shows it all. In Cell B1 is the raw score, in Cell B2 is the mean, in Cell B3 is the standard deviation, and in Cell B4 is the function that computes the z score of 1.30.

	A	B	C	D
		B4 ▼ fx =STANDARDIZE(B1,B2,B3)		
1	Raw Score	98		
2	Mean	95		
3	s	2.3		
4	z score for a raw score of 98	1.30		
5				
6				

Figure 8.6 Using the STANDARDIZE Function to Compute a z Score

But this can be a lot of work for a simple z score. So, let's apply it to a group of z scores. Take a look at Figure 8.7 (look familiar?). It contains the data we used for the first example of computing z scores earlier in the chapter. But here, we use the STANDARDIZE function for a set of scores, as well as the AVERAGE and STDEV functions, and copy the result down the column. Here's the formula we created from the STANDARDIZE, AVERAGE, and STDEV functions:

=STANDARDIZE(A5,AVERAGE($A$2:$A$11),STDEV($A$2:$A$11)).

And as you can see in Figure 8.7, it works!

	A	B	C	D	E	F	G	H	I	J
		B5 ▼ fx =STANDARDIZE(A5,AVERAGE($A$2:$A$11),STDEV($A$2:$A$11))								
1	X	z score								
2	12	0								
3	15	1.5								
4	11	-0.5								
5	13	0.5								
6	8	-2								
7	14	1								
8	12	0								
9	13	0.5								
10	12	0								
11	10	-1								
12										

Figure 8.7 Using the STANDARDIZE Function in a Formula to Compute the z Score for a Set of Raw Scores

What z Scores Represent

You already know that a particular z score represents a raw score but also represents a particular location along the x-axis of a distribution. And the more extreme the z score (such as -2 or $+2.6$), the further it is from the mean.

Because you already know the percentage of area that falls between certain points along the x-axis (such as 34% between the mean and a standard deviation of $+1$, or about 14% between a standard deviation of $+1$ and a standard deviation of $+2$), we can make the following statements that will be true as well:

- 84% of all the scores fall below a z score of $+1$ (the 50% that falls below the mean plus the 34% that falls between the mean and the $+1$ z score).
- 16% of all the scores fall above a z score of $+1$ (because the total area under the curve has to equal 100%, and 84% of the scores fall below a score of $+1.0$).

Think about both of these for a moment. All we are saying is that, given the normal distribution, different areas of the curve are encompassed by different numbers of standard deviations or z scores. OK—here it comes. These percentages or areas can also easily be seen as representing *probabilities* of a certain score occurring. For example, here's the big question (drum roll, please):

In a distribution with a mean of 100 and a standard deviation of 10, what is the probability that any one score will be 110 or above?

The answer? 16% or 16 out of 100 or .16. How did we get this?

First, we computed the corresponding z score, which is $+1$ [$(110 - 100)/10$]. Then, given the knowledge we already have (see Figure 8.4), we know a z score of 1 represents a location on the x-axis below which 84% (50% plus 34%) of all the scores in the distribution fall. Above that is 16% of the scores or a probability of .16. Because we already know the areas between the mean and 1, 2, or 3 standard deviations above or below the mean, we can easily figure out the probability that the value of any one z score has of occurring.

But the method we just went through is fine for z values of 1, 2, and 3. But what if the value of the z score is not a whole number like 2, but 1.23 or -2.01? We need to find a way to be more precise.

How do we do that? Simple—learn calculus and apply it to the curve to compute the area underneath it at almost every possible point along the x-axis, or (and we like this alternative much more) use Table B.1 found in Appendix B (the normal distribution table). This is a listing of all the values (except the most extreme) for the area under a curve that corresponds to different z scores. This table has two columns. The first column, labeled z score, is simply the z score that has been computed. The second column, *Area Between the Mean and the z Score*, is the exact area underneath the curve that is contained between the two points.

For example, if we wanted to know the area between the mean and a z score of +1, find the value 1.00 in the column labeled z score and read across to the second column where you find the area between the mean and a z score of 1.00 to be 34.13. Seen that before?

Why aren't there any plus or minus signs in this table (such as −1.00)? Because the curve is symmetrical and it does not matter if the value of the z score is positive or negative. The area between the mean and 1 standard deviation in any direction is always 34.13%.

Here's the next step. Let's say that for a particular z score of 1.38, you want to know the probability associated with that z score. If you wanted to know the percentage of the area between the mean and a z score of 1.38, you would find the corresponding area for the z score in Table B.1 of 1.38, which is 41.62, which indicates that more than 41% of all the cases in the distribution fall within a z score of 0 and 1.38, and that about 92% (50% plus 41.62%) will fall at or below a z score of 1.38. Now, you should notice that we did this last example without any raw scores at all. Once you get to this table, they are just no longer needed.

But are we always interested only in the amount of area between the mean and some other z score? What about between two z scores, neither of which is the mean? For example, what if we were interested in knowing the amount of area between a z score of 1.5 and a z score of 2.5, which translates to a probability that a score falls between the two z scores? How can we use the table to compute these outcomes? It's easy. Just find the corresponding amount of area each z score encompasses and subtract one from the other.

Often, drawing a picture helps, as in Figure 8.8.

For example, let's say that we want to find the area between a raw score of 110 and 125 in a distribution with a mean of 100 and a standard deviation of 10. Here are the steps we would take.

1. Compute the z score for a raw score of 110, which is (110 − 100)/10, or +1.

2. Compute the z score for a raw score of 125, which is (125 − 100)/10, or +2.5.

3. Using Table B.1 in Appendix B, find the area between the mean and a z score of +1, which is 34.13%.

4. Using Table B.1 in Appendix B, find the area between the mean and a z score of +2.5, which is 49.38%.

5. Because you want to know the distance between the two, subtract the smaller from the larger: 49.38 − 34.13, or 15.25%. Here's the picture that's worth a thousand words, in Figure 8.8.

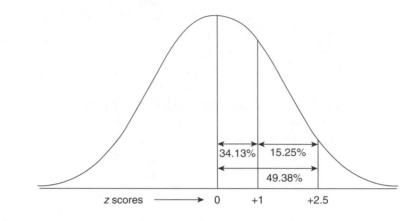

Figure 8.8   Using a Drawing to Figure Out the Difference in Area
Between Two z Scores

More Excel

There's another way to compute the probability associated with a particular raw score, and that's through the use of the **NORMSDIST** function. This takes the form of

=NORMSDIST(z),

where

z = the z score for which you want to find the probability.

So, if you want to know the probability associated with a z score of 1, then the function would look like this:

=NORMSDIST(1).

And the result that would be returned to the cell would be .841345, which should look quite familiar to you. Try it—it's a very cool way to get the values you need when Table B.1 is missing!

OK—so we can be pretty confident that the probability of a particular score occurring can be best understood by examining where that score falls in a distribution relative to other scores. In this example, the probability of a score occurring between a z score of +1 and a z score of +2.5 is about 15%.

Here's another example. In a set of scores with a mean of 100 and a standard deviation of 10, a raw score of 117 has a corresponding z score of 1.70. This z score corresponds to an area under the curve of 45.54%, which means that the probability of this score occurring between a score of 0 and a score of 1.70 is 95.54% (or 50% + 45.54%) or 95.5 out of 100 or .955.

TECH TALK Just two things about standard scores. First, even though we are focusing on z scores, there are other types of standard scores as well. For example, a *T* score is a type of standard score that is computed by multiplying the z score times 10 and adding 50. One advantage of this type of score is that you rarely have a negative *T* score. As with z scores, *T* scores allow you to compare standard scores from different distributions.

Second, a standard score is a whole different animal from a standardized score. A standardized score is one that comes from a distribution with a predefined mean and standard deviation. Standardized scores from tests such as the SAT and GRE (Graduate Record Exam) are used so that comparisons can be made easily between scores where the same mean and standard deviation are being used.

What z Scores Really Represent

The name of the statistics game is being able to estimate the probability of an outcome. If we take what we have talked about and done so far in this chapter one step further, it is deciding the probability of some event occurring. Then, we will use some criterion to judge whether we think that event is as likely, more likely, or less likely than what we would expect by chance. The research hypothesis presents a statement of the expected event, and we use our statistical tools to evaluate how likely that event is.

That's the 20-second version of what statistics is, but that's a lot. So let's take everything from this paragraph and go through it again with an example.

Let's say that your lifelong friend, trusty Lew, gives you a coin and asks you to determine if it is a "fair" one—that is, if you flip it 10 times you should come up with 5 heads and 5 tails. We would expect 5 heads (or 5 tails) because the probability is .5 of any one head or tail on any one flip. On 10 independent flips (meaning that one flip does not affect another), we should get 5 heads, and so on. Now the question is, how many heads would disqualify the coin as being fake or rigged?

Let's say the criterion for fairness that we will use is that if, in flipping the coin 10 times, we get heads (or heads turn up) less than 5% of the time, we'll say the coin is rigged and call the police on Lew (who, incidentally, is already on parole). This 5% is a standard that is used by statisticians. If the probability of the event (be it the number of heads or the score on a test or the difference between the average scores for two groups) occurs in the extreme (and we're saying the extreme is defined as less than 5% of all such occurrences), it's an unlikely, or in this case, an unfair, outcome.

Here's the distribution of how many heads you can expect, just by chance alone on 10 flips. All the possible combinations are 2^{10} or 1,024 possible outcomes, such as 9 heads and 1 tail, 7 heads and 3 tails, and 10 heads and 0 tails, and on and on. For example, the probability associated with getting 6 heads in 10 flips is about 21%.

Number of Heads	Probability
0	0.00
1	0.01
2	0.04
3	0.12
4	0.21
5	0.25
6	0.21
7	0.12
8	0.04
9	0.01
10	0.00

So, the likelihood of any particular outcome, such as 6 heads on 10 tosses, is about .21, or 21%. Now it's decision time. Just how many heads would one have to get on 10 flips to conclude that the coin is fixed, biased, busted, broken, or loony?

Well, as all good statisticians, we'll define the criterion as 5%, which we did earlier. If the probability of the observed outcome (the results of all our flips) is less than 5%, we'll conclude that it is so unlikely that something other than chance must be responsible—and our conclusion will be that the "something other than chance" is a bogus coin.

If you look at the table, you can see that 8, 9, or 10 heads all represent outcomes that are less than 5%. So if the result of 10 coin flips was 8, 9, or 10 heads, the conclusion would be that the coin is not a fair one. (Yep—you're right, 0, 1, and 2 qualify for the same decision. Sort of the other side of the coin—groan.)

The same logic applies to our discussion of z scores earlier. Just how extreme a z score would we expect before we could proclaim that an outcome is not due just to chance but to some other factor? If you look at the normal curve table in Appendix B, you'll see that the cutoff point for a z score of 1.65 includes about 45% of the area under the curve. If you add that to the other 50% of the area on the other side of the curve, you come up with a total of 95%. That leaves just 5% above that point on the x-axis. Any score that represents a z score of 1.65 or above is then into pretty thin air—or at least in a location that has a much smaller chance of occurring than others.

Hypothesis Testing and z Scores: The First Step

What we showed you here is that any event can have a probability associated with it. And we use those probability values to make decisions as to how unlikely we think an event might be. For example, it's highly unlikely to get only 1 head and 9 tails in 10 tosses of a coin. And we also said that if an event seems to occur only 5 out of 100 times (5%), we will deem that event to be rather unlikely relative to all the other events that could occur. It's much the same with any outcome related to a research hypothesis. The null hypothesis, which you learned about in Chapter 7, claims that there is no difference between groups (or variables) and that the likelihood of that occurring is 100%. We try to test the armor of the null for any chinks that might be there.

TECH TALK

z scores are very cool and useful, as are the corresponding *T* scores and *t* tests, which you will learn about in depth in Chapters 10 and 11. For now, though, it is important for you to know that *z* tests are reserved for populations and *t* tests are reserved for samples. This is why you so often see *t* tests rather than *z* tests reported in journal articles and other research reports.

In other words, if, through the test of the research hypothesis, we find that the likelihood of an event that occurred is somewhat extreme, then the research hypothesis is a more attractive explanation than would be the null. So, if we find a *z* score that is extreme (how extreme?—less than a 5% chance of occurring), we like to say that the reason for the extreme score is something to do with treatments or relationships and not just chance. We'll go into much greater detail on this point in the following chapter.

Summary

Being able to figure out a *z* score, and being able to estimate how likely it is to occur in a sample of data, is the first and most important skill in understanding the whole notion of inference. Once we know how likely a test score or a difference between groups is, we can compare that likelihood to what we would expect by chance and then make informed decisions. As we start Part IV of *Statistics for People Who (Think They) Hate Statistics . . . Excel 2007 Edition,* we'll apply this model to specific examples of testing questions about the difference.

Time to Practice

1. Normal curves:
   a. What are the characteristics of the normal curve?
   b. What human behavior, trait, or characteristic can you think of that is distributed normally?

2. Why is a z score a standard score, and why is it that z scores can be used to compare scores from different distributions with one another?

3. Compute the z scores for the following raw scores where $\bar{X} = 50$ and the standard deviation = 5.
   a. 55
   b. 50
   c. 60
   d. 58.5
   e. 46

4. Questions 4a through 4d are based on a distribution of scores with $\bar{X} = 75$ and the standard deviation = 6.38. Draw a small picture to help you see what's required.
   a. What is the probability of a score falling between a raw score of 70 and 80?
   b. What is the probability of a score falling above a raw score of 80?
   c. What is the probability of a score falling between a raw score of 81 and 83?
   d. What is the probability of a score falling below a raw score of 63?

5. Using the NORMSDIST function, compute the probability that a score would fall between +1 and +2 z scores. Remember, you don't need any raw score values because a z score of 1 is a z score of 1 is a z score of 1.

6. Jake needs to score in the top 10% in order to earn a physical fitness certificate. The class mean is 78, and the standard deviation is 5.5. What raw score does he need to get that valuable piece of paper?

7. So, why doesn't it make sense to simply combine, for example, course grades across different topics—just take an average and call it a day?

8. Who is the better student, relative to his or her classmates? Here's all the information you ever wanted to know.

Math			
Class Mean	81		
Class Standard Deviation	2		

Reading			
Class Mean	87		
Class Standard Deviation	10		

Raw Scores

	Math Score	Reading Score	Average
Noah	85	88	86.5
Talya	87	81	84

z Scores

	Math Score	Reading Score	Average
Noah			
Talya			

Answers to Practice Questions

1a. In a normal curve, the mean, median, and mode are equal to one another; the curve is symmetrical about the mean; and the tails are asymptotic.

1b. Height and weight are examples, as are intelligence and problem-solving skills.

2. A z score is a standard score (and comparable to others of the same type of score) because it is based on the degree of variability within its respective distribution of scores. Because a z score is always a measure of the distance between the mean and some point on the x-axis (regardless of the mean and standard deviation differences from one distribution to the next), the same units are used (units of standard deviations), and they can be compared to one another.

3a. $z = (55 - 50)/5 = +1.00$.

3b. $z = (50 - 50)/5 = 0$.

3c. $z = (60 - 50)/5 = +2.00$.

3d. $z = (58.5 - 50)/5 = +1.7$.

3e. $z = (46 - 50)/5 = -0.8$.

4a. The probability of a score falling between a raw score of 70 and a raw score of 80 is .5646. A z score for a raw score of 70 is $-.78$, and a z score for a raw score of 80 is .78. The area from the mean to a z score of .78 is 28.23%. The area between the two scores is 28.23 times 2, or 56.46%.

4b. The probability of a score falling above a raw score of 80 is .2167. A z score for a raw score of 80 is .78. The area between the mean and a z score of .78 is 28.23%. The area below a z score of .78 is .50 + .2833, or .7833. The difference between 1 (the total area under the curve) and .7833 is .2167.

4c. The probability of a score falling between a raw score of 81 and a raw score of 83 is .068. A z score for a raw score of 81 is .94, and a z score for a raw score of 83 is 1.25. The area from the mean to a z score of .94 is 32.64%. The area between the mean to a z score of 1.25 is 39.44%. The difference between the two is .3944 − .3264 = .068, or 6.8%.

4d. The probability of a score falling below a raw score of 63 is .03. A z score for a raw score of 63 is −1.88. The area between the mean and a z score of −1.88 is 46.99%. The area below a z score of 1.88 is 1 − (.50 + .4699) = .03.

5. This is easier than you think. Just enter the following formula, which uses the NORMSDIST function in any cell, and press enter:

$$=NORMSDIST(2)-NORMSDIST(1).$$

And the magic answer is 0.135905122.

6. Well, we know that the top 90% of the distribution is represented by a z of about 1.29, so we can plug in the values for the $X = z(s) + \bar{X}$, which is $X = 1.29(5.5) + 78$, which equals 85.095.

7. It doesn't because raw scores are not comparable to one another when they belong to different distributions. A raw score of 80 on the math test when the class mean was 40 is just not comparable with an 80 on the essay writing skills test where everyone got the one answer correct. Distributions, like people, are not always comparable to one another. Not everything (or everyone) is comparable to one another.

8. Here's the info with a few blank cells completed.

Math			
Class Mean	81		
Class Standard Deviation	2		
Reading			
Class Mean	87		
Class Standard Deviation	10		

Raw Scores			
	Math Score	Reading Score	Average
Noah	85	88	86.5
Talya	87	81	84

z Scores			
	Math Score	Reading Score	Average
Noah	2	0.1	1.05
Talya	3	−0.6	1.2

Noah has the higher average raw score (86.5 vs. 84 for Talya), but Talya has the higher average z score (1.2 vs. 1.05 for Noah). Remember that we asked who the better student was relative to the rest, which requires the use of a standard score (we used z scores). But why is Talya the better student relative to Noah? It's because on the test where there is the lowest variability (Math with an $sd = 2$), Talya really stands out with a z score of 3. That put her ahead to stay.

PART IV

Significantly Different

Using Inferential Statistics

Snapshots

"I see that David here is already busy computing *z* scores..."

Y ou've gotten this far and you're still alive and kicking, so congratulate yourself. By this point, you should have a good understanding of what descriptive statistics is all about, how chance figures as a factor in making decisions about outcomes, and how likely outcomes are to have occurred due to chance or some treatment.

You're an expert on creating and understanding the role that hypotheses play in social and behavioral science research. Now it's time for the rubber to meet the road. Let's see what you're made of in the next part of *Statistics for People Who (Think They) Hate Statistics . . . Excel 2007 Edition.* Best of all, the hard work you've put in will shortly pay off with an understanding of applied problems!

This part of the book deals exclusively with understanding and applying certain types of statistics to answer certain types of research questions. We'll cover the most common statistical tests, and even some that are a bit more sophisticated. At the end of this section, we'll show you some of the more useful software packages that can be used to compute the same values that we'll compute using a good old-fashioned calculator.

Let's start with a brief discussion of what the concept of significance is and go through the steps for performing an inferential test. Then we'll go on to examples of specific tests. We'll have lots of hands-on work here, so let's get started.

Significantly Significant

What It Means for You and Me

Difficulty Scale ☺☺ (somewhat thought provoking and key to it all!)

How much Excel? 📄 (just a mention)

What you'll learn about in this chapter

- What the concept of significance is and why it is important
- The importance of and difference between Type I and Type II errors
- How inferential statistics works
- How to select the appropriate statistical test for your purposes

THE CONCEPT OF SIGNIFICANCE

There is probably no term or concept that represents more confusion for the beginning statistics student than the concept of statistical significance. But that doesn't mean it has to be that way for you. Although it's a powerful idea, it is also relatively straightforward and can be understood by anyone in a basic statistics class.

We need an example of a study to illustrate the points we want to make. Let's take E. Duckett and M. Richards's "Maternal Employment and Young Adolescents' Daily Experiences in Single Mother Families" (paper presented at the Society for Research in Child Development, Kansas City, MO, 1989—a long time ago in a far-away galaxy . . .). These two authors examined the attitudes of 436 fifth- through ninth-grade adolescents toward maternal employment.

Specifically, the two researchers investigated whether differences are present between the attitudes of adolescents whose mothers work and the attitudes of adolescents whose mothers do not work. They also examined some other factors, but for this example, we'll stick with the mothers-who-work and mothers-who-don't-work groups. One more thing. Let's add the word *significant* to our discussion of differences, so we have a research hypothesis something like this:

> There is a significant difference in attitude toward maternal employment between adolescents whose mothers work and adolescents whose mothers do not work, as measured by a test of emotional state.

What we mean by the word *significant* is that any difference between the attitudes of the two groups is due to some systematic influence and not due to chance. In this example, that influence is whether or not mothers work. We assume that all of the other factors that might account for any differences between groups were controlled. Thus, the only thing left to account for the differences between adolescents' attitudes is whether or not mothers work. Right? Yes. Finished? Not quite.

If Only We Were Perfect

Because our world is not a perfect one, we must allow for some leeway in how confident we are that only those factors we identify could cause any difference between groups. In other words, you need to be able to say that although you are pretty sure the difference between the two groups of adolescents is due to maternal employment, you cannot be absolutely, 100%, positively, unequivocally, indisputably (get the picture?) sure. There's always a chance, no matter how small, that you are wrong.

Why? Many reasons. For example, you could (horrors) just be plain ol' wrong. Maybe during this one experiment, differences between adolescents' attitudes were not due to whether mothers worked or didn't work, but were due to some other factor that was inadvertently not accounted for, such as a speech given by the local Mothers Who Work Club that several students attended. How about if the people in one group were mostly adolescent males and the people in the other group were mostly adolescent females? That could be the source of a difference as well.

If you are a good researcher and do your homework, you can account for such differences, but it's always possible that you can't. And as a good researcher, you have to take that possibility into account.

So what do you do? In most scientific endeavors that involve testing hypotheses (such as the group differences example here), there is bound to be a certain amount of error that cannot be controlled—this is the chance factor that we have been talking about in the past few chapters. The level of chance or risk you are willing to take is expressed as a significance level, a term that unnecessarily strikes fear in the hearts of even strong men and women.

Significance level (here's the quick-and-dirty definition) is the risk associated with not being 100% confident that what you observe in an experiment is due to the treatment or what was being tested—in our example, whether or not mothers worked. If you read that significant findings occurred at the .05 level (or $p < .05$ in tech talk and what you regularly see in professional journals), the translation is that there is 1 chance in 20 (or .05 or 5%) that any differences found were not due to the hypothesized reason (whether mom works) but to some other, unknown reason or reasons. Your job is to reduce this likelihood as much as possible by removing all of the competing reasons for any differences that you observed. Because you cannot fully eliminate the likelihood (because no one can control every potential factor), you assign some level of probability and report your results with that caveat.

In sum (and in practice), the researcher defines a level of risk that he or she is willing to take. If the results fall within the region that says, "This could not have occurred by chance alone—something else is going on," the researcher knows that the null hypothesis (which states an equality) is not the most attractive explanation for the observed outcomes. Instead, the research hypothesis (that there is an inequality or a difference) is the favored explanation.

Let's take a look at another example, this one hypothetical.

A researcher is interested in seeing whether there is a difference in the academic achievement of children who participated in a preschool program and children who did not participate. The null hypothesis is that the two groups are equal to each other on some measure of achievement.

The research hypothesis is that the mean score for the group of children that participated in the program is higher than the mean score for the group of children that did not participate in the program. As a good researcher, your job is to show (as best you can—and no one is so perfect that he or she can account for everything) that any difference that exists between the two groups is due only to the effects of the preschool experience and no other factor or combination of factors. However, through a variety of techniques (that you'll learn about in your Stat II class!), you control or eliminate all the possible sources of difference, such as the influence of parents' education, number of children in the family, and so on. Once these other potential explanatory variables are removed, the only remaining alternative explanation for differences is the effect of the preschool experience itself.

But can you be absolutely (which is pretty darn) sure? No, you cannot. Why? First, because you can never be sure that you are testing a sample that identically reflects the profile of the population. And even if the sample perfectly represents the population, there are always other influences that might affect the outcome and that you inadvertently missed when designing the experiment. There's always the possibility of error.

By concluding that the differences in test scores are due to differences in treatment, you accept some risk. This degree of risk is, in effect (drum roll, please), the level of statistical significance at which you are willing to operate.

Statistical significance (here's the formal definition) is the degree of risk you are willing to take that you will reject a null hypothesis when it is actually true. For our example above, the null says that there is no difference between the two sample groups (remember, the null is always a statement of equality). In your data, however, you did find a difference. That is, given the evidence you have so far, group membership seems to have an effect on achievement scores. In reality, however, maybe there is no difference. If you reject the null you stated, you would be making an error. The risk you take in making this kind of error (or the level of significance) is also known as a Type I error.

The World's Most Important Table (for This Semester Only)

Here's what it all boils down to.

A null hypothesis can be true or false. Either there really is no difference between groups, or there really and truly is an inequality (such as the difference between two groups). But remember, you'll never know this true state because the null cannot be tested directly (remember that the null applies only to the population).

And, as a crackerjack statistician, you can choose to either reject or accept the null hypothesis. Right? These four different conditions create the table you see here in Table 9.1.

Let's look at each cell.

More About Table 9.1

Table 9.1 has four important cells that describe the relationship between the nature of the null (whether it's true or not) and your action (accept or reject the null hypothesis). As you can see, the null can be either true or false, and you can either reject or accept it.

The most important thing about understanding this table is the fact that the researcher never really knows the true nature of the null hypothesis and whether there really is or is not a difference between groups. Why? Because the population (which the null represents) is never tested directly. Why? Because it's impractical to do so, and that's why we have inferential statistics.

TABLE 9.1 Different Types of Errors

		Action You Take	
		Accept the Null Hypothesis	*Reject the Null Hypothesis*
True nature of the null hypothesis	*The null hypothesis is really true.*	**1** ☺ Bingo, you accepted a null when it is true and there is really no difference between groups.	**2** ☹ Oops—you made a Type I error and rejected a null hypothesis when there really is no difference between groups. Type I errors are also represented by the Greek letter alpha, or α.
	The null hypothesis is really false.	**3** ☹ Uh-oh—you made a Type II error and accepted a false null hypothesis. Type II errors are also represented by the Greek letter beta, or β.	**4** ☺ Good job, you rejected the null hypothesis when there really are differences between the two groups. This is also called power, or $1 - b$.

- ☺   Cell 1 in Table 9.1 represents a situation where the null hypothesis is really true (there's no difference between groups), and the researcher made the correct decision accepting it. No problem here. In our example, our results would show that there is no difference between the two groups of children, and we have acted correctly by accepting the null that there is no difference.
- ☹   Oops. Cell 2 represents a serious error. Here, we have rejected the null hypothesis (that there is no difference) when it is really true (and there is no difference). Even though there is no difference between the two groups of children, we will conclude there is and that's an error—clearly a boo-boo called a **Type I error**, also known as the level of significance.
- ☹   Uh-oh, another type of error. Cell 3 represents a serious error as well. Here, we have accepted the null hypothesis (that there is no difference) when it is really false (and, indeed, there is a difference). We have said that even though there is a difference between the two groups of children, we will conclude there is not—clearly a boo-boo, also known as a **Type II error**.
- ☺   Cell 4 in Table 9.1 represents a situation where the null hypothesis is really false and the researcher made the correct decision in rejecting it. No problem here. In our example, our results show that there is a difference between the two groups of children, and we have acted correctly by rejecting the null that states there is no difference.

TECH TALK   So, if .05 is good and .01 is even "better," why not set your Type I level of risk at .000001? For every good reason that you will be so rigorous in your rejection of false null hypotheses that you may reject a null when it is actually true. Such a stringent Type I error rate allows for little leeway—indeed, the research hypothesis might be true but the associated probability might be .015—still quite rare, but missed with the too-rigid Type I level of error.

Back to Type I Errors

Let's focus a bit more on Cell 2, where a Type I error was made, because this is the focus of our discussion.

This Type I error, or level of significance, has certain values associated with it that define the risk you are willing to take in any test of the null hypothesis. The conventional levels set are between .01 and .05.

For example, if the level of significance is .01, it means that on any one test of the null hypothesis, there is a 1% chance you will reject the null hypothesis when the null is true and conclude that there is a group difference when there really is no group difference at all.

If the level of significance is .05, it means that on any one test of the null hypothesis, there is a 5% chance you will reject it when the null is true (and conclude that there is a group difference) when there really is no group difference at all. Notice that the level of significance is associated with an independent test of the null, and it is not appropriate to say that "on 100 tests of the null hypothesis, I will make an error on only 5, or 5% of the time."

In a research report, statistical significance is usually represented as $p < .05$, read as "the probability of observing that outcome is less than .05," often expressed in a report or journal article simply as "significant at the .05 level."

TECH TALK With the introduction of fancy-schmancy software such as Excel that can do statistical analysis, there's no longer the worry about the imprecision of such statements as "$p < .05$" or "$p < .01$." $p < .05$ can mean anything from .000 to .049999, right? Instead, software such as Excel gives you the *exact* probability such as .013 or .158 of the risk you are willing to take that you will commit a Type I error. So, when you see in a research article the statement that "$p < .05$," it means that the value of p is equal to anything from .00 to .049999999999 (you get the picture). Likewise, when you see "$p > .05$" or "$p =$ n.s." (for nonsignificant), it means that the probability of rejecting a true null exceeds .05 and, in fact, can range from .0500001 to 1.00.

So, it's actually terrific when we know the exact probability of an outcome because we can measure more precisely the risk we are willing to take.

There is another kind of error you can make, which, along with the Type I error, is shown in Table 9.1. A Type II error (Cell 3 in the chart) occurs when you inadvertently accept a false null hypothesis.

TECH TALK When talking about the significance of a finding, you might hear the word **power** used. Power is a construct that has to do with how well a statistical test can detect and reject a null hypothesis when it is false. Mathematically, it's calculated by subtracting the value of the Type II error from 1. A more powerful test is always more desirable than a less powerful test, because the more powerful one lets you get to the heart of what's false and what's not.

For example, there may really be differences between the populations represented by the sample groups, but you mistakenly conclude there are not.

Ideally, you want to minimize both Type I and Type II errors, but it is not always easy or under your control. You have complete control over the Type I error level or the amount of risk that you are willing to take (because you actually set the level itself). Type II errors are not as directly controlled but, instead, are related to factors such as sample size. Type II errors are particularly sensitive to the number of subjects in a sample, and as that number increases, Type II error decreases. In other words, as the sample characteristics more closely match those of the population (achieved by increasing the sample size), the likelihood that you will accept a false null hypothesis decreases.

SIGNIFICANCE VERSUS MEANINGFULNESS

What an interesting situation for the researcher when discovering that the results of an experiment indeed are statistically significant. You know technically what statistical significance means—that the research was a technical success and the null hypothesis is not a reasonable explanation for what was observed. Now, if your experimental design and other considerations were well taken care of, statistically significant results are unquestionably the first step toward making a contribution to the literature in your field. However, the value of statistical significance and its importance or meaningfulness must be kept in perspective.

For example, let's take the case where a very large sample of illiterate adults (say, 10,000) is divided into two groups. One group receives intensive training to read using computers, and the other receives intensive training to read using classroom teaching. The average score for Group 1 (which learned in the classroom) is 75.6 on a reading test, the outcome variable. The average score on the reading test for Group 2 (which learned using the computer) is 75.7. The amount of variance in both groups is about equal. As you can see, the difference in score is only one tenth of 1 point (75.6 vs. 75.7), but when a *t* test for the significance between independent means is applied, the results are significant at the .01 level, indicating that computers do work better than classroom teaching. (The next two chapters discuss *t* tests.)

The difference of 0.1 is indeed statistically significant, but is it meaningful? Does the improvement in test scores (by such a small margin) provide sufficient rationale for the $300,000 it costs to fund the program? Or is the difference negligible enough that it can be ignored, even if it is statistically significant?

Here are some conclusions about the importance of statistical significance that we can reach given this and the countless other possible examples:

- Statistical significance, in and of itself, is not very meaningful unless the study that is conducted has a sound conceptual base that lends some meaning to the significance of the outcome.
- Statistical significance cannot be interpreted independently of the context within which it occurs. For example, if you are the superintendent in a school system, are you willing to retain children in Grade 1 if the retention program significantly raises their standardized test scores by one-half point?
- Although statistical significance is important as a concept, it is not the end-all and certainly should not be the only goal of scientific research. That is the reason why we set out to test hypotheses rather than prove them. If our study is designed correctly, then even null results tell you something very important. If a particular treatment does not work, it is important information that others need to know about. If your study is designed well, then you should know why the treatment does not work, and the next person down the line can design his or her study taking into account the valuable information you provided.

AN INTRODUCTION TO INFERENTIAL STATISTICS

Whereas descriptive statistics are used to describe a sample's characteristics, inferential statistics are used to infer something about the population based on the sample's characteristics.

At several points throughout the first half of *Statistics for People Who (Think They) Hate Statistics . . . Excel 2007 Edition*, we have emphasized that a hallmark of good scientific research is choosing a sample in such a way that it is representative of the population from which it was selected. The process then becomes an inferential one, where you infer from the smaller sample to the larger population based on the results of tests (and experiments) conducted using the sample.

Before we start discussing individual inferential tests, let's go through the logic of how the inferential method works.

How Inference Works

Here are the general steps of a research project to see how the process of inference might work. We'll stay with adolescents' attitudes toward mothers working as an example.

Here's the sequence of events that might happen:

1. The researcher selects representative samples of adolescents who have mothers who work and adolescents who have mothers who do not work. These are selected in such a way that the samples represent the populations from which they are drawn.

2. Each adolescent is administered a test to assess his or her attitude. The mean scores for groups are computed and compared using some test.

3. A conclusion is reached as to whether the difference between the scores is the result of chance (meaning some factor other than moms working is responsible for the difference), or the result of "true" and statistically significant differences between the two groups (meaning the results are due to moms working).

4. A conclusion is reached as to the relationship between maternal employment and adolescents' attitudes in the population from which the sample was originally drawn. In other words, an inference, based on the results of an analysis of the sample data, is made about the population of all adolescents.

How to Select What Test to Use

Step 3 above brings us to ask the question, "How do I select the appropriate statistical test to determine if a difference between groups exists?" Heaven knows, there are plenty of them (many hundreds), and you have to decide which one to use and when to use it. Well, the best way to learn which test to use is to be an experienced statistician who has taken lots of courses in this area and participated in lots of research. Experience is still the greatest teacher. In fact, there's no way you can really learn what to use and when to use it unless you've had the real-life, applied opportunity to actually use these tools. And as a result of taking this course, you are learning how to use these very tools.

So, for our purposes and to get started, we've created this nice little flow chart (a.k.a. cheat sheet) of sorts that you see in Figure 9.1. You have to have some idea of what you're doing, so selecting the correct statistical test is not entirely autopilot, but it certainly is a good place to get started.

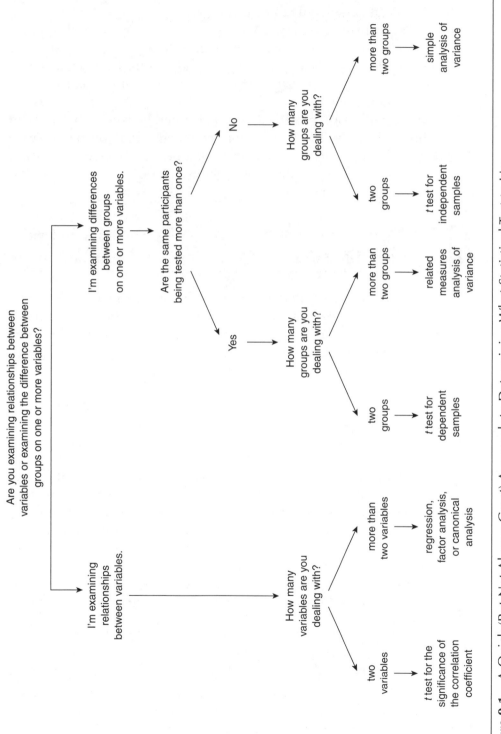

Figure 9.1 A Quick (But Not Always Great) Approach to Determining What Statistical Test to Use

Don't think for a second that Figure 9.1 takes the place of your need to learn about when these different tests are appropriate. The flow chart is here only to help you get started.

This is really important. We just wrote that selecting the appropriate statistical test is sure not an easy thing to do. And the best way to do it is to learn how to do it, and that means practicing and even taking more statistics courses. The simple flow chart we have here works, but use it with caution. When you make a decision, check with your professor or some other person who has been through this stuff and feels more confident than you might (who also knows more!). You can also find a neat tool, named Statistical Navigator, at http://rimarcik.com/en/navigator that you may find interesting and of some help.

Here's How to Use the Chart

1. Assume that you're very new to this statistics stuff (which you are) and that you have some idea of what these tests of significance are, but you're pretty lost as far as deciding which one to use when.

2. Answer the question at the top of the flow chart.

3. Proceed down the chart by answering each of the questions until you get to the end of the chart. That's the statistical test you should use. This is not rocket science, and with some practice (which you will get throughout this part of *Statistics for People . . .*), you'll be able to quickly and reliably select the appropriate test. Each of the chapters in this part of the book will begin with a chart like the one you see in Figure 9.1 and take you through the specific steps for the test statistic you should use.

Does the cute flow chart in Figure 9.1 contain all the statistical tests there are? Not by a long shot. There are hundreds, but the ones in Figure 9.1 are the ones used most often. And if you are going to become familiar with the research in your own field, you are bound to run into these.

AN INTRODUCTION TO TESTS OF SIGNIFICANCE

What inferential statistics does best is that it allows decisions to be made about populations based on the information about samples. One of the most useful tools for doing this is a test of statistical significance that can be applied to different types of situations depending on the nature of the question being asked and the form of the null hypothesis.

For example, do you want to look at the difference between two groups, such as whether boys score significantly differently from girls on some test? Or the relationship between two variables, such as number of children in a family and average score on intelligence tests? The two cases call for different approaches, but both will result in a test of a null hypothesis using a specific test of statistical significance.

How a Test of Significance Works: The Plan

Tests of significance are based on the fact that each type of null hypothesis has associated with it a particular type of statistic. And each of the statistics has associated with it a special distribution that you compare with the data you obtain from a sample. A comparison between the characteristics of your sample and the characteristics of the test distribution allows you to conclude if the sample characteristics are different from what you would expect by chance.

Here are the general steps to take in the application of a statistical test to any null hypothesis. These steps will serve as a model for each of the chapters that follow in Part IV.

1. A statement of the null hypothesis. Do you remember that the null hypothesis is a statement of equality? The null hypothesis is the "true" state of affairs given no other information on which to make a judgment.

2. Setting the level of risk (or the level of significance or Type I error) associated with the null hypothesis. With any research hypothesis comes a certain degree of risk that you are wrong. The smaller this error is (such as .01 compared with .05), the less risk you are willing to take. No test of a hypothesis is completely risk free because you never really know the "true" relationship between variables. Remember that it is traditional to set the Type I error rate at .01 or .05; Excel and other programs specify the exact level.

3. Selection of the appropriate test statistic. Each null hypothesis has associated with it a particular test statistic. You can learn what test is related to what type of question in this part of *Statistics for People* . . .

4. Computation of the test statistic value. The **test statistic value** (called the **obtained value**) is the result of a specific statistical test. For example, there are test statistics for the significance of the difference between the averages of two groups, for the significance of the difference of a correlation coefficient from 0, and for the significance of the difference between two proportions. You'll actually compute the test statistic and come up with a numerical value.

5. Determination of the value needed for rejection of the null hypothesis using the appropriate table of critical values for the particular statistic. Each test statistic (along with group size and the risk you are willing to take) has a **critical value** associated with it. This is the value you would expect the test statistic to yield if the null hypothesis is indeed true.

6. Comparison of the obtained value to the critical value. This is the crucial step. Here, the value you obtained from the test statistic (the one you computed) is compared with the value (the critical value) you would expect to find by chance alone.

7. If the obtained value is more extreme than the critical value, the null hypothesis cannot be accepted. That is, the null hypothesis's statement of equality (reflecting chance) is not the most attractive explanation for differences that were found. Here is where the real beauty of the inferential method shines through. Only if your obtained value is more extreme than chance (meaning that the result of the test statistic is not a result of some chance fluctuation) can you say that any differences you obtained are not due to chance and that the equality stated by the null hypothesis is not the most attractive

explanation for any differences you might have found. Instead, the differences must be due to the treatment.

8. If the obtained value does not exceed the critical value, the null hypothesis is the most attractive explanation. If you cannot show that the difference you obtained is due to something other than chance (such as the treatment), then the difference must be due to chance or something you have no control over. In other words, the null is the best explanation.

Here's the Picture That's Worth a Thousand Words

What you see in Figure 9.2 represents the eight steps that we just went through. This is a visual representation of what happens when the obtained and critical values are compared. In this example, the significance level is set at .05, or 5%. It could have been set at .01, or 1%.

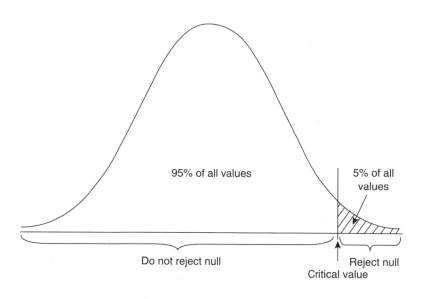

Figure 9.2   Comparing Obtained Values to Critical Values and Making Decisions About Rejecting or Accepting the Null Hypothesis

In examining Figure 9.2, note the following:

1. The entire curve represents all the possible outcomes based on a specific null hypothesis, such as the difference between two groups or the significance of a correlation coefficient.

2. The critical value is the point beyond which the obtained out-comes are judged to be so rare that the conclusion is that the obtained outcome is not due to chance but to some other fac-tor. In this example, we define rare as having a less than 5% chance of occurring.

3. If the outcome representing the obtained value falls to the left of the critical value (that is, is less extreme), the conclusion is that the null hypothesis is the most attractive explanation for any differences that are observed. In other words, the obtained value falls in the region (95% of the area under the curve) where we expect only outcomes due to chance to occur.

4. If the obtained value falls to the right of the critical value (it is more extreme), the conclusion is that the research hypothesis is the most attractive explanation for any differences that are observed. In other words, the obtained value falls in the region (5% of the area under the curve) where we would expect only outcomes due to something other than chance to occur.

Summary

So, now you know exactly how the concept of significance works, and all that is left is applying it to a variety of different research questions. That's what we'll start with in the next chapter and continue with through most of this part of the book.

Time to Practice

1. Why is significance an important construct in the study and use of inferential statistics?

2. What's wrong with the following statements?
    a. A Type I error of .05 means that in 5 tests out of 100 tests of the research hypothesis, I will reject a true null hypothesis.
    b. It is possible to set the Type I error rate to 0.
    c. The smaller the Type I error rate, the better the results.

3. What does chance have to do with testing the research hypothesis for significance?

4. What does the actual critical value represent?

5. In Figure 9.2 (page 215), there is a striped area in the right-hand part of the curve.
    a. What does that area represent?
    b. If you test the research hypothesis at a more rigorous level (say from .05 to .01), would that area get bigger or smaller, and why?

Answers to Practice Questions

1. The concept of significance is crucial to the study and use of inferential statistics because significance (reflected in the idea of a significance level) sets the level at which we can be confident that the outcomes we observe are "truthful" and to what extent these outcomes can be generalized to the larger population from which the sample was selected.

2a. Level of significance refers only to a single, independent test of the null hypothesis and not to multiple tests.

2b. No, no, and no. It is impossible to set the error rate to 0 because it is not possible that we might not reject a null hypothesis when it is actually true. There's always a chance.

2c. The level of risk that you are willing to take to reject the null hypothesis when it is true has nothing to do with the meaningfulness of the outcomes of your research. You can have a highly significant outcome that is meaningless, or have a relatively high Type I error rate (.10) and have a very meaningful finding.

3. Chance is reflected in the degree of risk (Type I error) that we are willing to take in the possible rejection of a true null hypothesis.

4. The critical value is the value that you would expect by chance alone (if the null were "correct"), and values more extreme indicate an outcome that is due to something other than chance. The critical value is a cutoff where less extreme values occur by chance (much) more than 5% of the time.

5a. That area under the curve represents outcomes that are so unlikely, we attribute them to something other than chance—perhaps the treatment.

5b. As the test becomes more rigorous, the area becomes smaller and the defining line where the area begins moves to the right. There's less room for error (of the Type I kind) and hence, the area is smaller.

10 *t*(ea) for Two

Tests Between the Means of Different Groups

Difficulty Scale ☺☺☺ (not too hard—this is the first one of this kind, but you know more than enough to master it)

How much Excel?  (lots)

What you'll learn about in this chapter

- When the *t* test for independent means is appropriate to use
- How to compute the observed *t* value
- How to use the TTEST function
- How to use the *t*-Test Analysis ToolPak tool for computing the *t* value
- Interpreting the *t* value and understanding what it means

INTRODUCTION TO THE *T* TEST FOR INDEPENDENT SAMPLES

Even though eating disorders are recognized for their seriousness, little research has been done that compares the prevalence and intensity of symptoms across different cultures. John P. Sjostedt, John F. Shumaker, and S. S. Nathawat undertook this comparison with groups of 297 Australian and 249 Indian university students. Each student was tested on the Eating Attitudes Test and the

Goldfarb Fear of Fat Scale. The groups' scores were compared with one another. On a comparison of means between the Indian and the Australian participants, Indian students scored higher on both of the tests. The results for the Eating Attitudes Test were $t_{(524)} = -4.19$, $p < .0001$, and the results for the Goldfarb Fear of Fat Scale were $t_{(524)} = -7.64$, $p < .0001$.

Now just what does all this mean? Read on.

Why was the t test for independent means used? Sjostedt and his colleagues were interested in finding out if there was a difference on the average scores of one (or more) variable(s) between the two groups that were independent of one another. By independent, we mean that the two groups were not related in any way. Each participant in the study was tested only once. The researchers applied a t test for independent means and arrived at the conclusion that for each of the outcome variables, the differences between the two groups were significant at or beyond the .0001 level. Such a small Type I error means that there is very little chance that the difference in scores between the two groups was due to something other than group membership, in this case representing nationality, culture, or ethnicity.

Want to know more? Check out Sjostedt, J. P., Shumaker, J. F., & Nathawat, S. S. (1998). Eating disorders among Indian and Australian university students. *Journal of Social Psychology, 138*(3), 351–357.

The Path to Wisdom and Knowledge

Here's how you can use Figure 10.1, the flow chart introduced in Chapter 10, to select the appropriate test statistic, the t test for independent means. Follow along the highlighted sequence of steps in Figure 10.1.

1. The differences between the groups of Australian and Indian students are being explored.

2. Participants are being tested only once.

3. There are two groups.

4. The appropriate test statistic is t test for independent means.

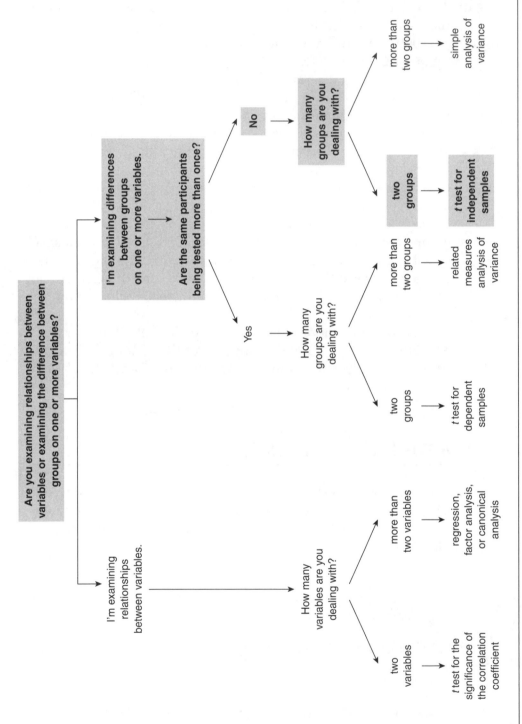

Figure 10.1 Determining That a *t* Test Is the Correct Statistic

TECH TALK

Almost every statistical test has certain assumptions that underlie the use of the test. For example, the t test has a major assumption that the amount of variability in each of the two groups is equal. This is the homogeneity of variance assumption. Although this assumption can be safely violated if the sample size is big enough, small samples and a violation of this assumption can lead to ambiguous results and conclusions. Don't knock yourself out worrying about these assumptions because they are beyond the scope of this book. However, you should know that such assumptions are rarely violated, but it is worth knowing that they do exist.

COMPUTING THE TEST STATISTIC

The formula for computing the t value for the t test for independent means is shown in Formula 10.1. The difference between the means makes up the numerator of the following formula used to compute the t value or the test statistic of the obtained value. The amount of variation within and between each of the two groups makes up the denominator.

$$t = \frac{\bar{X}_1 - \bar{X}_2}{\sqrt{\left[\frac{(n_1 - 1)s_1^2 + (n_2 - 1)s_2^2}{n_1 + n_2 - 2}\right]\left[\frac{n_1 + n_2}{n_1 n_2}\right]}}, \qquad (10.1)$$

where

$\bar{X}_1$ is the mean for Group 1,

$\bar{X}_2$ is the mean for Group 2,

n_1 is the number of participants in Group 1,

n_2 is the number of participants in Group 2,

s_1^2 is the variance for Group 1, and

s_2^2 is the variance for Group 2.

Nothing new here at all. It's just a matter of plugging in the correct values.

Here are some data reflecting the number of words remembered following a program designed to help Alzheimer's patients remember the order of daily tasks. Group 1 was taught using visuals, and

Group 2 was taught using visuals and intense verbal rehearsal. We'll use the data to compute the test statistic in the following example.

Group 1			Group 2		
7	5	5	5	3	4
3	4	7	4	2	3
3	6	1	4	5	2
2	10	9	5	4	7
3	10	2	5	4	6
8	5	5	7	6	2
8	1	2	8	7	8
5	1	12	8	7	9
8	4	15	9	5	7
5	3	4	8	6	6

Here are the famous eight steps and the computation of the t-test statistic.

1. A statement of the null and research hypotheses.

As represented by Formula 10.2, the null hypothesis states that there is no difference between the means for Group 1 and Group 2. For our purposes, the research hypothesis (shown as Formula 10.3) states that there is a difference between the means of the two groups. The research hypothesis is a two-tailed, nondirectional research hypothesis because it posits a difference, but in no particular direction. The null hypothesis is

$$H_0: \mu_1 = \mu_2. \tag{10.2}$$

The research hypothesis is

$$H_1: \bar{X}_1 \neq \bar{X}_2. \tag{10.3}$$

2. Setting the level of risk (or the level of significance or Type I error) associated with the null hypothesis.

The level of risk or Type I error or level of significance (any other names?) is .05, totally the decision of the researcher.

3. Selection of the appropriate test statistic.

Using the flow chart shown in Figure 10.1, we determined that the appropriate test is a *t* test for independent means. It is not a *t* test for dependent means (a common mistake beginning students make) because the groups are independent of one another.

4. Computation of the test statistic value (called the obtained value).

Now's your chance to plug in values and do some computation. The formula for the *t* value was shown in Formula 10.1. When the specific values are plugged in, we get the equation shown in Formula 10.4. (We already computed the mean and standard deviation.)

$$t = \frac{5.43 - 5.53}{\sqrt{\left[\frac{(30-1) \times 3.42^2 + (30-1) \times 2.06^2}{30+30-2}\right]\left[\frac{30+30}{30 \times 30}\right]}} \cdot \quad (10.4)$$

With the numbers plugged in, Formula 10.5 shows how we got the final value of −.1371. The value is negative because a larger value (the mean of Group 2, which is 5.53) is being subtracted from a smaller number (the mean of Group 1, which is 5.43).

Remember, though, that because the test is nondirectional and any difference is hypothesized, the sign of the difference is meaningless. In other words, you can ignore it!

$$t = \frac{-.1}{\sqrt{\left[\frac{339.20 + 123.06}{58}\right]\left[\frac{60}{900}\right]}} = -.14. \quad (10.5)$$

TECH TALK When a nondirectional test is discussed, you may find that the *t* value is represented as an absolute value looking like this, $|t|$, which ignores the sign of the value altogether. Your teacher may even express the *t* value as such to emphasize that the sign is relevant for a one-directional test, but surely not for a directional one.

---

5. Determination of the value needed for rejection of the null hypothesis using the appropriate table of critical values for the particular statistic.

Here's where we go to Table B.2 in Appendix B, which lists the critical values for the *t* test.

We can use this distribution to see if two independent means differ from one another by comparing what we would expect by chance (the tabled or critical value) to what we observe (the obtained value).

Our first task is to determine the **degrees of freedom** (*df*), which approximates the sample size. For this particular test statistic, the degrees of freedom are $n_1 - 1 + n_2 - 1$, or $n_1 + n_2 - 2$. So for each group, add the size of the two samples and subtract 2. In this example, $30 + 30 - 2 = 58$. These are the degrees of freedom for this test statistic and not necessarily for any other.

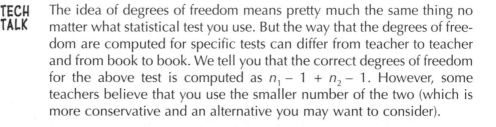

TECH TALK The idea of degrees of freedom means pretty much the same thing no matter what statistical test you use. But the way that the degrees of freedom are computed for specific tests can differ from teacher to teacher and from book to book. We tell you that the correct degrees of freedom for the above test is computed as $n_1 - 1 + n_2 - 1$. However, some teachers believe that you use the smaller number of the two (which is more conservative and an alternative you may want to consider).

Using this number (58), the level of risk you are willing to take (earlier defined as .05), and a two-tailed test (because there is no direction to the research hypothesis), you can use the *t*-test table to look up the critical value. At the .05 level, with 58 degrees of freedom for a two-tailed test, the value needed for rejection of the null hypothesis is . . . oops! There's no 58 degrees of freedom in the table! What do you do? Well, if you select the value that corresponds to 55, you're being conservative in that you are using a value for a sample smaller than what you have (and the critical *t* value will be larger).

If you go for 60 degrees of freedom (the closest to your value of 58), you will be closer to the size of the population, but a bit liberal in that 60 is larger than 58. Although statisticians differ in their viewpoint as to what to do in this situation, let's always go with the value that's closest to the actual sample size. So, the value needed to reject the null hypothesis with 58 degrees of freedom at the .05 level of significance is 2.001.

6. A comparison of the obtained value and the critical value.

The obtained value is −.14, and the critical value for rejection of the null hypothesis that Group 1 and Group 2 performed differently is 2.001. The critical value of 2.001 represents the value at which chance is the most attractive explanation for any of the observed differences between the two groups given 30 participants in each group and the willingness to take a .05 level of risk.

7. and 8. Decision time!

Now comes our decision. If the obtained value is more extreme than the critical value (remember Figure 9.2), the null hypothesis cannot be accepted. If the obtained value does not exceed the critical value, the null hypothesis is the most attractive explanation.

In this case, the obtained value (−.14) does not exceed the critical value (2.001)—it is not extreme enough for us to say that the difference between Groups 1 and 2 occurred by anything other than chance. If the value were greater than 2.001, it would represent a value that is just like getting 8, 9, or 10 heads in a coin toss—too extreme for us to believe that something else other than chance is not going on. In the case of the coin, it's an unfair coin—in this example, it would be that there is a better way to teach memory skills to these older people.

So, to what can we attribute the small difference between the two groups? If we stick with our current argument, then we could say the difference is due to anything from sampling error to rounding error to simple variability in participants' scores. Most important, we're pretty sure (but, of course, not 100% sure) that the difference is not due to anything in particular that one group or the other experienced to make its scores better.

So How Do I Interpret $t_{(58)} = -.14, p > .05$?

- *t* represents the test statistic that was used.
- 58 is the number of degrees of freedom.
- −.14 is the obtained value, obtained using the formula we showed you earlier in the chapter.
- $p > .05$ (the really important part of this little phrase) indicates that the probability is greater than 5% that on any one test of the null hypothesis, the two groups differ because of the way they were taught. Also note that the $p > .05$ can also appear as $p = $ n.s. for nonsignificant.

And Now . . . Using Excel's TTEST Function

Interestingly, Excel does not have a function that computes the *t* value for the difference between two independent groups. Rather, **TTEST** returns the probability of that value occurring. Very useful, but if you need the *t* value for a report, you may be out of luck (actually, not on your life—the Analysis ToolPak has a nifty one). Here are the steps.

1. Enter the individual scores into one column in a worksheet, and label one as Group 1 and one as Group 2 as you see in Figure 10.2.

	A	B
1	Group 1	Group 2
2	7	5
3	3	4
4	3	4
5	2	5
6	3	5
7	8	7
8	8	8
9	5	8
10	8	9
11	5	8
12	5	3
13	4	2
14	6	5
15	10	4
16	10	4
17	5	6
18	1	7
19	1	7
20	4	5
21	3	6
22	5	4
23	7	3
24	1	2
25	9	7
26	2	6
27	5	2
28	2	8
29	12	9
30	15	7
31	4	6

Figure 10.2   Data for Using the TTEST Function

2. Select the cell into which you want to enter the TTEST function.

In this example, we are going to have the TTEST value returned to Cell E5 (and that location was not chosen for any particular reason).

3. Now use the Formulas tab → More Functions → Statistical menu option and scroll down to select TTEST.

The function looks like this:

$$= TTEST(array1, array2, tails, type),$$

where

array1 = the cell addresses for the first set of data (which in this case is A2:A31);

array2 = the cell addresses for the second set of data (which in this case is B2:B31);

tails = 1 or 2 depending on whether this is a one-tailed (directional, which is a 1) or two-tailed (nondirectional, which
is a 2) test; and

type = 1 if it is a paired *t* test, 2 if it is a two-sample test (independent with equal variances), and 3 if it is a two-sample
test with unequal variances.

4. For this example, (which you will see in Figure 10.3 below), the finished function TTEST looks like this:

$$= TTEST(A2:A31,B2:B31,2,1).$$

Click OK, and you see the value returned, which is 0.877992, as you
see in Figure 10.3.

	E7			f_x	=TTEST(A2:A31,B2:B31,2,1)	
	A	B	C	D	E	
1	Group 1	Group 2				
2	7	5				
3	3	4				
4	3	4				
5	2	5				
6	3	5				
7	8	7			0.877991509	

Figure 10.3   Using TTEST to Compute the Probability of a
                 t Value

Notice two important things about TTEST.

It does not compute the *t* value.

It returns the likelihood that the resulting *t* value is due to
chance. As we said earlier, the interpretation of a *t* value with an
associated probability of .88 (and remember, it can only go to 1 or
100% likely) is pretty darn high.

Believe it or not, way back in the olden days, when your author
and perhaps your instructor were graduate students, there were
only huge mainframe computers and not a hint of such marvels as
we have today on our desktops. In other words, everything that
was done in our statistics class was done only by hand. The great
benefit of that is, first, it helps you to better understand the process.
Second, should you be without a computer, you can still do the
analysis. So, if the computer does not spit out all of what you need,
use some creativity. As long as you know the basic formula for the
obtained value and have the appropriate tables, you'll do fine.

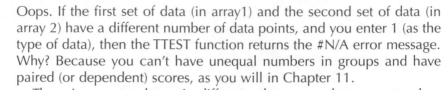

TECH TALK

Oops. If the first set of data (in array1) and the second set of data (in array 2) have a different number of data points, and you enter 1 (as the type of data), then the TTEST function returns the #N/A error message. Why? Because you can't have unequal numbers in groups and have paired (or dependent) scores, as you will in Chapter 11.

There is a pretty dramatic difference between what you get when you compute the t value using the formula and when you use one of the functions. Remember that when you did it manually, you had to use a table to locate the critical value and then compare the observed value to that? Well, with TTEST and the Analysis ToolPak discussion that follows later in this chapter, there's no more "$p <$." That's because Excel computes the exact, exactamento, precise, one-of-a-kind probability. No need for tables—just get that number, which is the probability associated with the outcome.

USING THE AMAZING ANALYSIS TOOLPAK TO COMPUTE THE *T* VALUE

Once again, we'll find that the ToolPak gives us all the information we need to make a very informed judgment about the value of *t* and its significance. The ToolPak tool also provides us with other information that, as you will see, is very helpful and saves us the effort of extra analyses as well.

1. Click Data → Data Analysis, and you will see the Data Analysis dialog box shown in Figure 10.4.

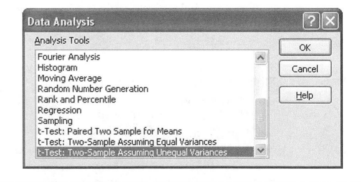

Figure 10.4   The Dialog Box That Gets Us Started With the Analysis ToolPak

2. Click t-test: Two-Sample Assuming Equal Variances and then click OK, and you will see the Descriptive Statistics dialog box as shown in Figure 10.5.

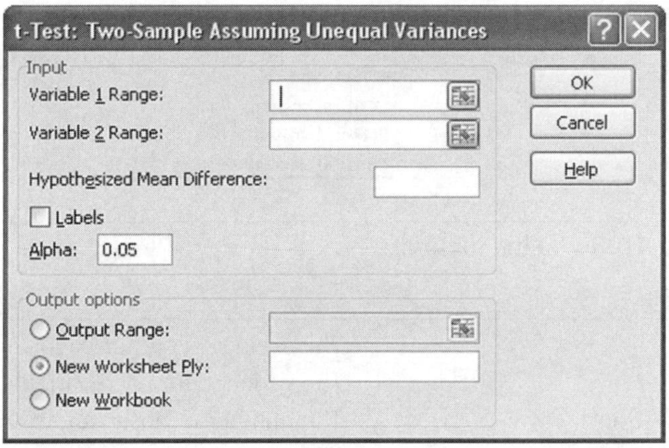

Figure 10.5   The t-Test Dialog Box

3. In the Variable 1 Range, enter the cell addresses for the first group of data. In our sample spreadsheet that you saw in Figure 10.2, the cell addresses are A1:A31 (and this includes the label Group 1).

4. In the Variable 2 Range, enter the cell addresses for the second group of data. In our sample spreadsheet that you saw in Figure 10.2, the cell addresses are B1:B31 (and this includes the label Group 2).

5. Click the Labels box so that labels are included.

6. Click the Output Range button, and enter an address where you want the output located on the same worksheet as the data. In this example, we are placing the output beginning in Cell D2.

7. Click OK, and as you can see in Figure 10.6, you get a tidy summary of important data relating to this analysis, including the following output and what it means.

	A	B	C	D	E	F
1	Group 1	Group 2		t-Test: Two-Sample Assuming Unequal Variances		
2	7	5				
3	3	4			Group 1	Group 2
4	3	4		Mean	5.43	5.53
5	2	5		Variance	11.70	4.26
6	3	5		Observations	30	30
7	8	7		Hypothesized Mean Difference	0.00	
8	8	8		df	48.00	
9	5	8		t Stat	-0.14	
10	8	9		P(T<=t) one-tail	0.45	
11	5	8		t Critical one-tail	1.68	
12	5	3		P(T<=t) two-tail	0.89	
13	4	2		t Critical two-tail	2.01	

Figure 10.6   Data Summary

Mean	The average score for each variable
Variance	The variance for each variable
Observations	The number of observations in each group
Pooled Variance	The variance for both groups
Hypothesized Mean Difference	What you may have indicated to be the difference you expect (back in the dialog box)
df	The degrees of freedom
t Stat	The value of the *t* statistic
P(T<=t) one-tail	The probability of *t* occurring by chance for a one-tailed test
t Critical one-tail	The critical value one needs to exceed for a one-tailed test (Remember those critical values from Chapter 8?)
P(T<=t) two-tail	The probability of *t* occurring by chance for a two-tailed test
t Critical two-tail	The critical value one needs to exceed for a two-tailed test (Remember those critical values from Chapter 8?)

More Excel

Remember that it takes only a moment to pretty up the ToolPak table, to copy it from Excel, and to paste it (or whatever you need from it) into another document.

Results

The results of the analysis showed that although Group 2 did have a higher score than Group 1, that score was not significantly different. The *t* value for a two-tailed test was −.14, with an associated *p* value of .89.

	Group 1	Group 2
Mean	5.43	5.53
Variance	11.70	4.26
Observations	30	30
Degrees of freedom	58.00	
t statistic	-0.11	
p value	0.89	

SPECIAL EFFECTS: ARE THOSE DIFFERENCES FOR REAL?

OK, now you have some idea how to test for the difference between the averages of two separate or independent groups. Good job. But that's not the whole story.

You may have a significant difference between groups, but the $64,000 question is not only whether that difference is (statistically) significant, but also whether it is *meaningful*. We mean, is there enough of a separation between the distribution that represents each group so that the difference you observe and the difference you test is really a difference? Hmm. . . . Welcome to the world of effect size. **Effect size** is a measure of how different two groups are from one another—it's a measure of the magnitude of the treatment, kind of like how big is big. And what's especially interesting about computing effect size is that sample size is not taken into account.

Calculating effect size, and making a judgment about it, adds a whole new dimension to understanding significant outcomes.

Let's take the following example. A researcher tests the question of whether participation in community-sponsored services (such as card games, field trips, etc.) increases the quality of life (as rated from 1 to 10) for older Americans. The researcher implements the treatment over a 6-month period and then, at the end of the treatment period, measures quality of life in the two groups (each consisting of 50 participants over the age of 80 where one group got the services and one group did not). Here are the results.

	No Community Services	Community Services
Mean	7.46	6.90
Standard deviation	1.03	1.53

And, the verdict is that the difference is significant at the .034 level (which is $p < .05$, right?).

OK, there's a significant difference, but what about the *magnitude* of the difference?

The great Pooh-bah of effect size was Jacob Cohen, who wrote some of the most influential and important articles on this topic. He authored a very important and influential book (your stat teacher has it on his or her shelf) that instructs researchers how to figure out the effect size for a variety of different questions that are asked about differences and relationships between variables. Here's how you do it.

Computing and Understanding the Effect Size

Just as with many other statistical techniques, there are many different ways to compute the effect size. We are going to show you the most simple and straightforward. You can learn more about effect sizes at some of the references we'll be giving you in a minute.

By far, the most direct and simple way to compute effect size is to simply divide the difference between the means by any one of the standard deviations. Danger, Will Robinson—this does assume that the standard deviations (and the amount of variance) between groups are equal to one another. For our example above, we'll do this:

$$ES = \frac{\bar{X}_1 - \bar{X}_2}{s}, \tag{10.6}$$

where

  ES = effect size,

  $\bar{X}_1$ = the mean for Group 1,

  $\bar{X}_2$ = the mean for Group 2, and

  s = the standard deviation from either group.

So, in our example,

$$ES = \frac{7.46 - 6.90}{1.53}, \tag{10.7}$$

or .366. So, the effect size for this example is .37.

What does it mean? One of the very cool things that Cohen (and others) figured out was just what a small, medium, and large effect size is. They used the following guidelines:

A small effect size ranges from 0.0 to .20.

A medium effect size ranges from .20 to .50.

A large effect size is any value above .50.

Our example, with an effect size of .37, is categorized as medium. But what does it really mean?

Effect size gives us an idea about the relative positions of one group to another. For example, if the effect size is 0, that means that both groups tend to be very similar and overlap entirely—there is no difference between the two distributions of scores. On the other hand, an effect size of 1 means that the two groups overlap about 45% (having that much in common). And, as you might expect, as the effect size gets larger, it reflects the increasing lack of overlap between the two groups.

Jacob Cohen's book, *Statistical Power Analysis for the Behavioral Sciences,* first published in 1967 with the latest edition (1988) available from Lawrence Erlbaum Associates (which was recently purchased by Taylor and Francis, so that's where you might have to go to find it), is a must for anyone who wants to go beyond the very general information that is presented here. It is full of tables and techniques for allowing you to understand how a statistically significant finding is only half the story—the other half is the magnitude of that effect.

TECH TALK

So, you really want to be cool about this effect size thing. You can do it the simple way, as we just showed you (by subtracting means from one another and dividing by either standard deviation), or you can really wow that good-looking classmate who sits next to you. The grown-up formula for the effect size uses the pooled variance in the numerator of the ES equation that you saw above. The pooled standard deviation is sort of an average of the standard deviation from Group 1 and the standard deviation from Group 2. Here's the formula:

$$ES = \frac{\bar{X}_1 - \bar{X}_2}{\sqrt{\dfrac{\sigma_1^2 + \sigma_2^2}{2}}}, \tag{10.8}$$

where

ES = effect size,

$\bar{X}_1$ = the mean of Group 1,

$\overline{X}_2$ = the mean of Group 2,

σ_1^2 = the variance of Group 1, and

σ_2^2 = the variance of Group 2.

If we applied this formula to the same numbers we showed you above, you'd get a whopping effect size of .43—not very different from .37, which we got using the more direct method shown earlier (and still in the same category of medium size). But it is a more precise method, and one that is well worth knowing about.

A Very Cool Effect Size Calculator

Why not take the A train and just go right to http://web.uccs .edu/lbecker/Psy590/escalc3.htm, where statistician Lee Becker from the University of Colorado at Colorado Springs developed an effect size calculator? With this calculator, you just plug in the values, click Compute, and the program does the rest, as you see in Figure 10.7. Thanks, Dr. Becker!

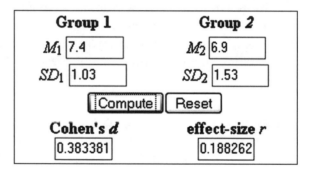

Figure 10.7　The Very Cool Effect Size Calculator

Summary

The *t* test is your first introduction to performing a real statistical test and trying to understand this whole matter of significance from an applied point of view. Be sure that you understand what was in this chapter before you move on. And be sure you can do by hand the few things that were asked for. Next, we move on to using another form of the same test, only this time, there are two measures taken from one group of participants rather than one measure taken from two separate groups.

Time to Practice

1. Using the data in the file named Chapter 10 Data Set 1, test the research hypothesis at the .05 level of significance that boys raise their hands in class more often than girls. Do this practice problem by hand using a calculator. What is your conclusion regarding the research hypothesis? Remember to first decide whether this is a one- or two-tailed test.

2. Using the same data set (Chapter 10 Data Set 1), test the research hypothesis at the .01 level of significance that there is a difference between boys and girls in the number of times they raise their hands in class. Do this practice problem by hand using a calculator. What is your conclusion regarding the research hypothesis? You used the same data for this problem as for Question 1, but you have a different hypothesis (one is directional and the other is nondirectional). How do the results differ, and why?

3. Using the data in the file named Chapter 10 Data Set 2, test the null hypothesis that urban and rural residents both have the same attitude toward gun control. Use the TTEST function to test the hypothesis.

4. For your Friday afternoon report to the boss, you need to let her know if the two stores in Newark, Delaware, are selling at the same weekly rate or a different rate. Use the data in the file named Chapter 10 Data Set 3 and the Analysis ToolPak to let her know. Better hurry.

Answers to Practice Questions

1. The mean for boys equals 7.93, and the mean for girls equals 5.31. The obtained *t* value is 3.006, and the critical *t* value at the .05 level for rejection of the null hypothesis for a one-tailed test (boys *more* than girls, remember?) is 1.701. Conclusion? Boys raise their hands significantly more!

2. Now this is very interesting. We have the same exact data, of course, but a different hypothesis. Here, the hypothesis is that the number of times is *different* (not just more or less), necessitating a two-tailed test. So, using Table B.2 and at the .01 level for a two-tailed test, the critical value is 2.764. The obtained value of 3.006 (same results as when you did the analysis for #1 above) does exceed what we would expect by chance, and given this hypothesis, there is a difference. So, in comparison with one another, a one-tailed finding (see Question 1 above) need not be as extreme as a two-tailed finding given the same data to reach the same conclusion (that the research hypothesis is supported).

3. We used the TTEST function and got a value of .254. No muss, no fuss—that's the exact probability that these two groups of scores are

different as a function of residence. And, guess what? They absolutely don't differ from one another.

4. And the results of the two-store comparison are (and remember, we made it appear a bit prettier than the plain output).

	Main Street	Mall
Mean	$4309.47	$2608.80
Variance	1844069.981	1598119.457
Observations	15	15
df	28	
t Stat	3.550156066	
P(T<=t) one-tail	0.000691671	
P(T<=t) two-tail	0.001383343	

Because the associated probability with the difference of $1700.67 (4309.47 – 2608.80) is .0013 (and remember, it is two-tailed, because the boss only wants to know about the difference and not what direction), obviously they are different, and a major review of the mall store (and its employees) has to take place!

11

t(ea) for Two (Again)

Tests Between the Means of Related Groups

Difficulty Scale ☺☺☺ (hard—just like the one in Chapter 10, but a different question)

How much Excel? 🗷 🗷 (some)

What you'll learn about in this chapter

- When the *t* test for dependent means is appropriate to use
- How to compute the observed *t* value
- Interpreting the *t* value and understanding what it means
- How to use the TTEST function and the TDIST functions
- How to use the t-Test ToolPak tool for computing the *t* value

INTRODUCTION TO THE *T* TEST FOR DEPENDENT SAMPLES

How best to educate children is clearly one of the most vexing questions that faces any society. Because children are so different from one another, a balance needs to be found between meeting the basic needs of all while ensuring that special children (on either end of the continuum) get the opportunities they need. An obvious and important part of education is reading, and three professors at the University of Alabama studied the effects of resource and regular classrooms on the reading achievement of learning-disabled children. Renitta Goldman, Gary L. Sapp, and Ann Shumate Foster found that, in general, 1 year of daily instruction in both settings resulted in no difference in overall reading achievement scores. On one specific comparison between the pretest and the posttest scores of the resource group, they found that $t_{(34)} = 1.23$, $p > .05$. At the beginning of the program, reading achievement scores for children in the resource room were 85.8. At the end of the program, reading achievement scores for children in the resource room were 88.5—a difference, but not a significant one.

So why a test of dependent means? A t test for dependent means indicates that a single group of the same subjects is being studied under two conditions. In this example, the conditions are before the start of the experiment and after its conclusion. Primarily, it is because the same children were tested at two times, before the start of the 1-year program and at the end of the 1-year program, that we use the t test for dependent means. As you can see by the above result, there was no difference in scores at the beginning and the end of the program. The very small t value (1.23) is not nearly extreme enough to fall outside the region where we would reject the null hypothesis. In other words, there is far too little change for us to say that this difference occurred by something other than chance. The small difference of 2.7 (88.5 – 85.8) is probably due to sampling error or variability within the groups.

Want to know more? Check out Goldman, R., Sapp, G. L., & Foster, A. S. (1998). Reading achievement by learning disabled students in resource and regular classes. *Perceptual and Motor Skills, 86,* 192–194.

The Path to Wisdom and Knowledge

Here's how you can use the flow chart to select the appropriate test statistic, the t test for dependent means. Follow along the highlighted sequence of steps in Figure 11.1.

 TECH TALK There's another way that statisticians talk of dependent tests—as repeated measures. They are also called this because the measures are repeated across time or conditions or some factor, but they are repeated across the same cases, be the case a person, place, or thing.

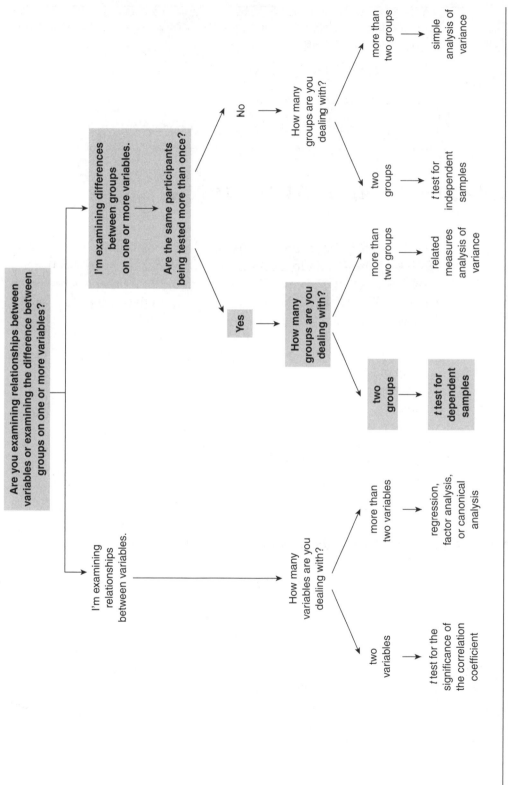

Figure 11.1 Determining That a *t* Test for Dependent Means Is the Correct Test Statistic

1. The difference between the students' scores on the pretest and on the posttest is the focus.

2. Participants are being tested more than once.

3. There are two groups.

4. The appropriate test statistic is *t* test for dependent means.

COMPUTING THE TEST STATISTIC

The *t* test for dependent means involves a comparison of means from each group and focuses on the differences between the scores. As you can see in Formula 11.1, the sum of the differences between the two tests forms the numerator and reflects the difference between groups.

$$t = \frac{\Sigma D}{\sqrt{\dfrac{n\Sigma D^2 - (\Sigma D)^2}{n-1}}}, \tag{11.1}$$

where

ΣD is the sum of all the differences between groups,

ΣD^2 is the sum of the differences squared between groups, and

n is the number of pairs of observations.

Here are some data to illustrate how the *t*-value is computed. Just like in the above example, there is a pretest and a posttest, and for illustration's sake, assume that these are before and after scores from a reading program.

Pretest	Posttest	Difference	D^2
3	7	4	16
5	8	3	9
4	6	2	4
6	7	1	1
5	8	3	9
5	9	4	16
4	6	2	4

	Pretest	Posttest	Difference	D^2
	5	6	1	1
	3	7	4	16
	6	8	2	4
	7	8	1	1
	8	7	−1	1
	7	9	2	4
	6	10	4	16
	7	9	2	4
	8	9	1	1
	8	8	0	0
	9	8	−1	1
	9	4	−5	25
	8	4	−4	16
	7	5	−2	4
	7	6	−1	1
	6	9	3	9
	7	8	1	1
	8	12	4	16
Sum	158	188	30	180
Mean	6.32	7.52	1.2	7.2

Here are the famous eight steps and the computation of the *t*-test statistic.

1. A statement of the null and research hypotheses.

The null hypothesis states that there is no difference between the means for the pretest and the posttest scores on reading achievement. The research hypothesis is a one-tailed, directional research hypothesis because it posits that the posttest score will be higher than the pretest score.
The null hypothesis is

$$H_0: \mu_{postest} = \mu_{pretest}. \tag{11.2}$$

The research hypothesis is

$$H_1: \bar{X}_{postest} > \bar{X}_{pretest}. \tag{11.3}$$

2. Setting the level of risk (or the level of significance or Type I error) associated with the null hypothesis.

The level of risk or Type I error or level of significance is .05, totally the decision of the researcher.

3. Selection of the appropriate test statistic.

Using the flow chart shown in Figure 11.1, we determined that the appropriate test is a t test for dependent means. It is not a t test for independent means because the groups are not independent of each other. In fact, they're not groups of participants, but groups of scores for the same participants. The groups (and the scores) are dependent on one another. Another name for the t test for dependent means is the t test for paired samples or the t test for correlated samples. You'll see in Chapter 14 that there is a very close relationship between a test of the significance of the correlation between these two sets of scores (pre and post) and the t value we are computing here.

4. Computation of the test statistic value (called the obtained value).

Now's your chance to plug in values and do some computation. The formula for the t value was shown above. When the specific values are plugged in, we get the equation shown in Formula 11.4. (We already computed the means and standard deviations for the pretest and posttest scores.)

$$t = \frac{30}{\sqrt{\dfrac{(25 \times 180) - 30^2}{25 - 1}}}. \tag{11.4}$$

With the numbers plugged in, we have the following equation with a final obtained t value of 2.45. The mean score for pretest performance was 6.32, and the mean score for posttest performance was 7.52.

$$t = \frac{30}{\sqrt{150}} = 2.45. \tag{11.5}$$

5. Determination of the value needed for rejection of the null hypothesis using the appropriate table of critical values for the particular statistic.

Here's where we go to Table B.2, which lists the critical values for the t test. Once again, we have a t test, and we'll use the same table we used in Chapter 10 to find out the critical value for rejection of the null hypothesis.

Our first task is to determine the degrees of freedom (*df*), which approximate the sample size. For this particular test statistic, the degrees of freedom are $n - 1$ where *n* equals the number of pairs of observations, or $25 - 1 = 24$. These are the degrees of freedom for this test statistic only and not necessarily for any other.

Using this number (24), the level of risk you are willing to take (earlier defined as .05), and a one-tailed test (because there is a direction to the research hypothesis—the posttest score will be larger than the pretest score), the value needed for rejection of the null hypothesis is 1.711.

6. A comparison of the obtained value and the critical value is made.

The obtained value is 2.45, larger than the critical value needed for rejection of the null hypothesis.

7. and 8. Time for a decision.

Now comes our decision. If the obtained value is more extreme than the critical value, the null hypothesis cannot be accepted. If the obtained value does not exceed the critical value, the null hypothesis is the most attractive explanation. In this case, the obtained value does exceed the critical value—it is extreme enough for us to say that the difference between the pretest and the posttest did occur by something other than chance. And if we did our experiment correctly, then what could the factor be that affected the outcome? Easy—the introduction of the daily reading program. We know the difference is due to a particular factor. The difference between the pretest and the posttest groups could not have occurred by chance, but instead is due to the treatment.

So How Do I Interpret $t_{(24)} = 2.45$, p < .05?

- *t* represents the test statistic that was used.
- 24 is the number of degrees of freedom.
- 2.45 is the obtained value using the formula we showed you earlier in the chapter.
- $p < .05$ (the really important part of this little phrase) indicates that the probability is less than 5% on any one test of the null hypothesis that the average of posttest scores is greater than the average of pretest scores due to chance alone—there's something else going on. Because we defined .05 as our criterion for the research hypothesis being more attractive than the null hypothesis, our conclusion is that there is a significant difference between the two sets of scores. That's the something else.

And Now . . . Using Excel's TTEST Function

Interestingly, Excel (like the independent test we covered in Chapter 10) uses the same function that computes the *t* value for the difference between two dependent groups. The TTEST function returns the probability of that value occurring, and you use it the same exact way as we showed you in Chapter 10 (only you change one of the options, as you will see).

More Excel

In addition to the TTEST function, there's the **TDIST** function where you enter the value of *t*, the degrees of freedom, and the number of tails (1 or 2) at which the hypothesis is being tested, and Excel returns the probability of the outcome. So, for example, remember your normal curve teachings in Chapter 8? A *t* of 1.96 with *df* = 10,000 (virtually identical to a *z* score) will have a probability of about .05. Aren't you smart!

Enter the individual scores into two columns in a worksheet, and label one as Pretest and one as Posttest, as you see in Figure 11.2 (and the same data we used in the earlier example).

	A	B
1	Pretest	Postest
2	3	7
3	5	8
4	4	6
5	6	7
6	5	8
7	5	9
8	4	6
9	5	6
10	3	7
11	6	8
12	7	8
13	8	7
14	7	9
15	6	10
16	7	9
17	8	9
18	8	8
19	9	8
20	9	4
21	8	4
22	7	5
23	7	6
24	6	9
25	7	8
26	8	12

Figure 11.2   Data for Using the TTEST Function With a Set of Dependent Means

Select the cell into which you want to enter the TTEST function. In this example, we are going to have the TTEST value returned to Cell D5 (and that location was not chosen for any particular reason).

Now use the Formulas → More Functions → Statistical → TTEST menu options and the "Inserting a Function" technique we talked about on page XXX in Chapter 1 to enter the TTEST function in Cell D5. The function looks like this:

=TTEST(array1,array2,tails,type),

where

array1 = the cell addresses for the first set of data (which in this case is A2:A26);

array2 = the cell addresses for the second set of data (which in this case is B2:B26);

tails = 1 or 2 depending on whether this is a one-tailed (directional, which is a 1) or two-tailed (nondirectional, which is a 2) test, and in this case, it is one tailed, so we enter a 1; and

type = 1 or 2 depending on whether the variances are equal (1) or not equal (2), and in this example, the variances are equal, so we enter a 1.

For this example, (which you will see in Figure 11.3 below), the finished function TTEST looks like this (and note that the type is equal to 1 because the observations are paired or dependent):

=TTEST(A2:A26,B2:B26,1,1).

Click OK and you see the value returned, which is 0.021983, as you see in Figure 11.3.

As with the conclusion we reached earlier, the treatment did have a significant effect because the probability of the difference between the pretest and posttest being due to chance was less than .021983 or 2%. Pretty rare.

D5				f_x	=TTEST(A2:A26,B2:B26,2,1)	
	A	B	C	D	E	F
1	Pretest	Postest				
2	3	7				
3	5	8				
4	4	6				
5	6	7		0.021983		

Figure 11.3 Using TTEST to Compute the Probability of a *t* Value

Once again, remember two important things about the TTEST function.

- It does not, does not, does not (this is an easy mistake to make) compute the *t* value.
- It returns the likelihood that the resulting *t* value is due to chance. For example, the interpretation of a *t* value with an associated probability of .88 (and remember, it can go only up to 1, or 100% likely) is pretty darn high.

USING THE AMAZING ANALYSIS TOOLPAK TO COMPUTE THE *T* VALUE

Once again, as with the *t* test for the difference between independent means, the ToolPak gives us the tools and all the information we need to make a very informed judgment about the value of *t* and its significance. Here we go, with a very similar procedure to what we did in Chapter 10, only this time, we are using the same data and selecting the t-Test: Paired Two Sample for Means option.

1. Click Data → Data Analysis, and you will see the Data Analysis dialog box shown in Figure 11.4.

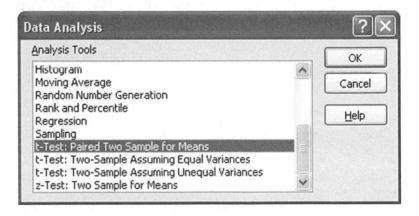

Figure 11.4   The Analysis ToolPak Dialog Box That Gets us Started

2. Click the t-Test: Paired Two Sample for Means (Excel's way of describing dependent means) option, then click OK and you will see the Descriptive Statistics dialog box as shown in Figure 11.5.

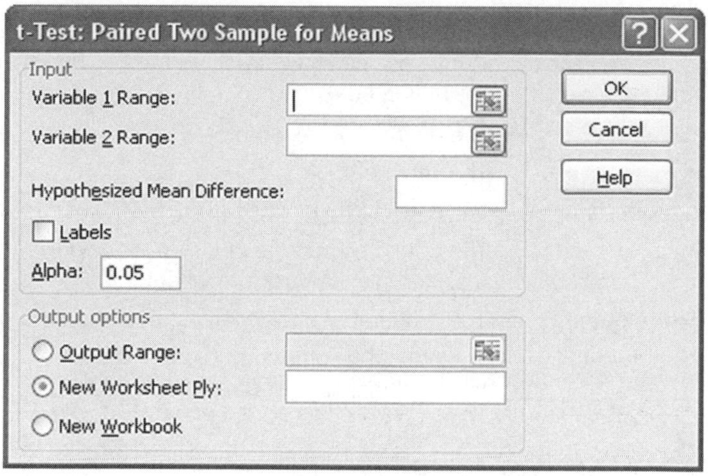

Figure 11.5   The *t* Test for Dependent Means or Paired
Samples Dialog Box

3. In the Variable 1 Range, enter the cell addresses for the pretest.
   In our sample spreadsheet that you saw in Figure 11.2, the cell
   addresses are A1:A26 (and this includes the label Pretest).

4. In the Variable 2 Range, enter the cell addresses for the posttest.
   In our sample spreadsheet that you saw in Figure 11.2, the cell
   addresses are B1:B26 (and this includes the label Posttest).

5. Click the Labels box so that labels are included.

6. Click the Output Range button and enter an address where you
   want the output located on the same worksheet as the data. In
   this example, we are placing the output beginning in Cell D2.

7. Click OK, and as you can see in Figure 11.6, you get a tidy
   summary of important data relating to this analysis, including
   the following output and what it means.

Mean	The average score for each variable
Variance	The variance for each variable
Observations	The number of observations in each group, which will always be the same because they are "paired" and there are two observations for each case.
Pearson Correlation	The degree of relationship between both variables (see Chapter 5 to refresh your memory about this)

(Continued)

(Continued)

Hypothesized Mean Difference	What you may have indicated to be the difference you expect (back in the dialog box)
df	The degrees of freedom
t Stat	The value of the *t* statistic
P(T<=t) one-tail	The probability of *t* occurring by chance for a one-tailed test
t Critical one-tail	The critical value one needs to exceed for a one-tailed test (remember those critical values from Chapter 9?)
P(T<=t) two-tail	The probability of *t* occurring by chance for a two-tailed test
t Critical two-tail	The critical value one needs to exceed for a two-tailed test (remember those critical values from Chapter 9?)

	A	B	C	D	E	F
1	Pretest	Postest				
2	3	7		t-Test: Paired Two Sample for Means		
3	5	8				
4	4	6			Pretest	Postest
5	6	7		Mean	6.32	7.52
6	5	8		Variance	2.98	3.34
7	5	9		Observations	25.00	25.00
8	4	6		Pearson Correlation	0.05	
9	5	6		Hypothesized Mean Difference	0.00	
10	3	7		df	24.00	
11	6	8		t Stat	-2.45	
12	7	8		P(T<=t) one-tail	0.01	
13	8	7		t Critical one-tail	1.71	
14	7	9		P(T<=t) two-tail	0.02	
15	6	10		t Critical two-tail	2.06	

Figure 11.6   The Output for the ToolPak Analysis of a *t* Test Between Paired Samples or a *t* Test Between Dependent Means

Once again, our conclusion from earlier is supported. The probability of a *t* value of 2.45 occurring by chance alone is a bit more than .02, which is pretty tiny. Our conclusion? Must be something else going on, and what that something else is, is the implementation of a treatment.

 TECH TALK   So what's with the minus sign for the *t* Stat value in Figure 11.6? The only reason it's there is the way that the ToolPak computes the *t* value. It always subtracts the second value from the first, and because the second value is larger, it results in a negative sign. If you identified the Posttest as the first array, the value of –2.45 would have appeared as 2.45, but other values would not have changed. What's important is the significance of the *t* value (which in this case, for a one-tailed test, is about .02).

Summary

That's it for two group designs that use means. You've just learned how to compare data from independent (Chapter 10) and dependent (Chapter 11) groups, and now it's time to move on to another class of significance tests that deals with more than two groups (be they independent or dependent). This class of techniques, called analysis of variance, is very powerful and popular and will be a valuable tool in your war chest!

Time to Practice

1. What is the difference between a test of independent means and a test of dependent means, and when is each appropriate?

2. For Chapter 11 Data Set 1, compute the t value using the ToolPak and write a conclusion as to whether there was a change in tons of paper used as a function of the recycling program in 25 different districts. (Hint: before and after become the two levels of treatment.) Test the hypothesis at the .01 level.

3. For Chapter 11 Data Set 2, compute the t value (do it manually, argh!), and write a conclusion as to whether there is a difference in satisfaction level in a group of families' use of service centers following a social service intervention. Then do this exercise using Excel, and report the exact probability of the outcome.

4. Ace used-car salesman Jack had no idea whether the $100,000 he pays for training for his used car salespeople works. Even Dave, the leading salesperson, thought it worked, but couldn't really tell. So, they hired your local helpful stat star and asked whether the training made a difference. Here are the data—you tell Jack whether it worked.

Month	Before Training	After Training
December	$58,676	$87,890
January	$46,567	$87,876
February	$87,656	$56,768
March	$65,431	$98,980
April	$56,543	$98,784
May	$45,456	$65,414
June	$67,656	$99,878
July	$78,887	$67,578
August	$65,454	$76,599
September	$56,554	$88,767
October	$58,876	$78,778
November	$54,433	$98,898

Answers to Practice Questions

1. A *t* test for independent means tests two distinct and different groups of participants, and each group is tested once. A *t* test for dependent means tests one group of participants, and each participant is tested twice.

2. The mean for before the recycling program was 34.44, and the mean for after was 34.8. There is an increase in recycling. Is the difference across the 25 districts significant? The obtained *t* value is .234, and with 24 degrees of freedom, the difference is not significant at the .01 level (in fact, the probability is .408)—the level at which the research hypothesis is being tested. Conclusion: The recycling program does not result in an increase in paper recycled.

3. There was an increase in level of satisfaction, from 5.48 to 7.60, which results in a *t* value of 3.893. This difference has an associated probability level of .001. It's very likely that the social service intervention worked (pretty well, in fact).

4. Yes for Jack! The *t* value is 3.34, which is significant at the .003 level. The average sales before the training was $61,849.08, and afterward, it was $83,850.83. Smart Jack.

12

Two Groups Too Many?

Try Analysis of Variance

Difficulty Scale ☺☺☺ (longer and harder than the others, but a very interesting and useful procedure—worth the work!)

How much Excel? ▣ ▣ ▣ ▣ (lots and lots)

What you'll learn about in this chapter

- What analysis of variance is and when it is appropriate to use
- How to compute and interpret the *F* statistic
- Using the FTEST and FDIST functions
- How to use the ANOVA: Single Factor ToolPak tool for computing the *F* value

INTRODUCTION TO ANALYSIS OF VARIANCE

One of the upcoming fields in the area of psychology is the psychology of sports. Although the field focuses mostly on enhancing performance, many aspects of sports receive special attention. One aspect focuses on what psychological skills are necessary to be a successful athlete. With this question in mind, Marious Goudas, Yiannis Theodorakis, and Georgios Karamousalidis have tested the usefulness of the Athletic Coping Skills Inventory.

As part of their research, they used a simple **analysis of variance** (or ANOVA—practice saying this; it sounds very cool) to test the hypothesis that number of years of experience in sports is related to coping skill (or an athlete's score on the Athletic Coping Skills Inventory). ANOVA was used because more than two groups were being tested, and these groups were compared on their average performance. In particular, Group 1 included athletes with 6 years of experience or less, Group 2 included athletes with 7 to 10 years of experience, and Group 3 included athletes with more than 10 years of experience.

The test statistic for ANOVA is the *F* test (named for R. A. Fisher, the creator of the statistic), and the results showed that $F_{(2, 110)}$ = 13.08, *p* < .01. The means of the three groups did differ from one another in their score on the Peaking Under Pressure subscale of the test. In other words, any difference in test score is due to number of years of experience in athletics rather than some chance occurrence of scores.

Want to know more? Check out the original reference: Goudas, M., Theodorakis, Y., & Karamousalidis, G. (1998). Psychological skills in basketball: Preliminary study for development of a Greek form of the Athletic Coping Skills Inventory. *Perceptual and Motor Skills, 86,* 59–65.

The Path to Wisdom and Knowledge

Here's how you can use the flow chart shown in Figure 12.1 to select ANOVA as the appropriate test statistic. Follow along the highlighted sequence of steps.

1. We are testing for differences between scores of the different groups, in this case, the difference between the peaking scores of athletes.

2. The athletes are not being tested more than once.

3. There are three groups (6 years or less, 7–10 years, and more than 10 years of experience).

4. The appropriate test statistic is simple analysis of variance.

Different Flavors of ANOVA

ANOVA comes in many different flavors. The simplest kind, and the focus of this chapter, is the **simple analysis of variance,** where there is one factor or one treatment variable (such as group membership) being explored, and there are more than two levels within this factor. Simple ANOVA is also called **one-way analysis of variance** or **single factor** because there is only one grouping dimension. The technique is called analysis of variance because the variance due to differences in performance is separated into variance that's due to differences between individuals within groups and variance due to differences between groups. Then, the two types of variance are compared with one another.

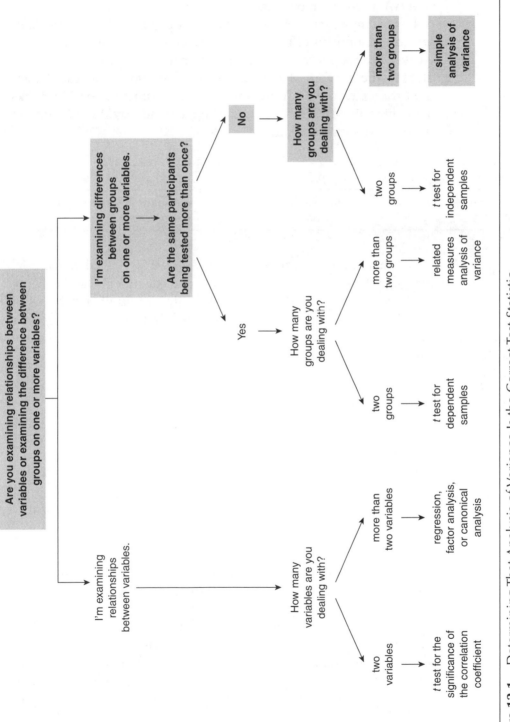

Figure 12.1 Determining That Analysis of Variance Is the Correct Test Statistic

In fact, ANOVA is, in many ways, similar to a *t* test. In both procedures, differences between means are computed. But with ANOVA, there are more than two means.

For example, let's say we were investigating the effects on language development of being in preschool for 5, 10, or 20 hours per week. The group to which the children belong is the treatment variable, or the grouping factor. Language development is the dependent variable, or the outcome. The experimental design looks something like this. It's a single factor or one-way ANOVA design like we mentioned earlier.

Group 1 (5 hours per week)	Group 2 (10 hours per week)	Group 3 (20 hours per week)
Language development test score	Language development test score	Language development test score

Another, more complex type of ANOVA is called a **factorial design,** where there is more than one treatment factor being explored. Here's an example where the effect of number of hours of preschool participation is being examined, but the effects of gender differences are being examined as well. The experimental design can look something like this:

	Number of Hours of Preschool Participation		
Gender	Group 1 (5 hours per week)	Group 2 (10 hours per week)	Group 3 (20 hours per week)
Male	Language development test score	Language development test score	Language development test score
Female	Language development test score	Language development test score	Language development test score

This factorial design is described as a 3 × 2 factorial design. The 3 indicates that there are three levels of one grouping factor (Group 1, Group 2, and Group 3). The 2 indicates that there are two levels of the other grouping factor (male and female). In combination, there are six different possibilities (males who spend 5 hours per week in preschool, females who spend 5 hours per week in preschool, males who spend 10 hours per week in preschool, etc.).

These factorial designs follow the same basic logic and principles of simple ANOVA, but they are just more ambitious in that they can test the influence of more than one factor at a time as well as a combination of factors. Don't worry—you'll learn all about factorial designs in the next chapter.

COMPUTING THE F TEST STATISTIC

Simple ANOVA involves testing the difference between the means of more than two groups on one factor or dimension. For example, you might want to know whether four groups of people (20, 25, 30, and 35 years of age) differ in their attitude toward public support of private schools. Or, you might be interested in determining whether five groups of children from different grades (2nd, 4th, 6th, 8th, and 10th) differ in the level of parental participation in school activities.
Any analysis where

- there is only one dimension or treatment,
- there are more than two levels of the grouping factor, and
- one is looking at differences across groups in average scores

requires that simple ANOVA be used.

The formula for the computation of the F value, which is the test statistic needed to evaluate the hypothesis that there are overall differences between groups, is shown in Formula 12.1. It is simple at this level, but it takes a bit more effort to compute than some of the other test statistics with which you have worked in earlier chapters.

$$F = \frac{MeanSquares_{Between}}{MeanSquares_{Within}}. \qquad (12.1)$$

TECH TALK The logic behind this ratio goes something like this. If there was absolutely no variability within each group (all the scores were the same), then any difference between groups would be meaningful, right? Probably so. The ANOVA formula (which is a ratio) compares the amount of variability between groups (which is due to the grouping factor) to the amount of variability within groups (which is due to chance). If that ratio is 1, then the amount of variability due to within-group differences is equal to the amount of variability due to between-group differences, and any difference between groups would not be significant. As the average difference between groups gets larger (and the numerator of the ratio increases in value), the F value increases as well. As the F value increases, it becomes more extreme in relation to the distribution of all F values and is more likely due to something other than chance. Whew!

Here are some data and some preliminary calculations to illustrate how the F value is computed. For our example, let's assume these are three groups of preschoolers and their language scores.

Group 1 Scores	Group 2 Scores	Group 3 Scores
87	87	89
86	85	91
76	99	96
56	85	87
78	79	89
98	81	90
77	82	89
66	78	96
75	85	96
67	91	93

Here are the famous eight steps and the computation of the *F*-test statistic.

1. A statement of the null and research hypotheses.

The null hypothesis, shown in Formula 12.2, states that there is no difference between the means for the three different groups. ANOVA, also called the *F* test (because it produces an *F* statistic or an *F* ratio or an *F* value), looks for an overall difference between groups.

$$H_0: \mu_1 = \mu_2 = \mu_3.$$
(12.2)

It does not look at pairwise differences, such as the difference between Group 1 and Group 2. For that, we have to use another technique, which we discuss later in the chapter.

The research hypothesis, shown in Formula 12.3, states that there is an overall difference among the means of the three groups. Note that there is no direction to the difference because all *F* tests are nondirectional.

$$H_1: \bar{X}_1 \neq \bar{X}_2 \neq \bar{X}_3.$$
(12.3)

Up to now, we've talked about one- and two-tailed tests. No such thing when talking about ANOVA. Because more than two levels of a treatment factor are being tested, and because the *F* test is an omnibus (how's that for a word?) test (meaning that it tests for an overall difference between means), talking about the direction of specific differences does not make any sense.

2. Setting the level of risk (or the level of significance or Type I error) associated with the null hypothesis.

The level of risk or Type I error or level of significance (any other names?) is .05. Once again, the level of significance used is totally at the discretion of the researcher.

3. Selection of the appropriate test statistic.

Using the flow chart shown in Figure 12.1, we determined that the appropriate test is a simple ANOVA.

4. Computation of the test statistic value (called the obtained value).

Now's your chance to plug in values and do some computation. There's a good deal of computation to do.

- The F ratio is a ratio of variability between groups to variability within groups. To compute these values, we first have to compute what is called the sum of squares for each source of variability—between groups, within groups, and the total.
- The between-group sum of squares is equal to the sum of the differences between the mean of all scores and the mean of each group's score, which is then squared. This gives us an idea of how different each group's mean is from the overall mean.
- The within-group sum of squares is equal to the sum of the differences between each individual score in a group and the mean of each group, which is then squared. This gives us an idea of how different each score in a group is from the mean of that group.
- The total sum of squares is equal to the sum of the between-group and within-group sum of squares. OK, let's figure these values.

Figure 12.2 shows the practice data you saw above with all the calculations you need to compute the between-group, within-group, and total sum of squares.

First, let's look at what we have in this expanded table. Starting down the left column:

n	is the number of participants in each group (such as 10),
$\sum X$	is the sum of the scores in each group (such as 766),
$\overline{X}$	is the mean of each group (such as 76.60),

Group	Test Score	X^2	Group	Test Score	X^2	Group	Test Score	X^2
1	87	7,569	2	87	7,569	3	89	7,921
1	86	7,396	2	85	7,225	3	91	8,281
1	76	5,776	2	99	9,801	3	96	9,216
1	56	3,136	2	85	7,225	3	87	7,569
1	78	6,084	2	79	6,241	3	89	7,921
1	98	9,604	2	81	6,561	3	90	8,100
1	77	5,929	2	82	6,724	3	89	7,921
1	66	4,356	2	78	6,084	3	96	9,216
1	75	5,625	2	85	7,225	3	96	9,216
1	67	4,489	2	91	8,281	3	93	8,649
n	10		10			10		
ΣX	766			852			916	
$\overline{X}$	76.60			85.20			91.60	
$\Sigma(X^2)$	59,964			72,936			84,010	
$(\Sigma X)^2/n$	58,675.60			72,590.40			83,905.60	

$$N = 30.00$$
$$\Sigma\Sigma X = 2{,}534.00$$
$$(\Sigma\Sigma X)^2/N = \mathbf{214{,}038.53}$$
$$\Sigma\Sigma(X^2) = \mathbf{216{,}910}$$
$$\Sigma(\Sigma X)^2/n = \mathbf{215{,}171.60}$$

Figure 12.2   Computing the Important Values for a One-Way ANOVA

$\Sigma(X^2)$ is the sum of each score squared (such as 59,964), and

$(\Sigma X)^2/n$ is the sum of the scores in each group squared and then divided by the size of the group (such as 58,675.60).

Second, let's look at the right-most column:

N is the total number of participants (such as 30),

$\Sigma\Sigma X$ is the sum of all the scores across groups,

$(\Sigma\Sigma X)^2/N$ is the sum of all the scores across groups squared and divided by N,

$\Sigma\Sigma(X^2)$ is the sum of all the sums of squared scores, and

$\Sigma(\Sigma\Sigma X)^2/n$ is the sum of the sum of each group's scores squared and divided by n.

That is a load of computation to carry out, and we are almost finished.

First, we compute the sum of squares for each source of variability. Here are the calculations:

Between sum of squares	$\Sigma(\Sigma X)^2/n - (\Sigma\Sigma X)^2/N$, or $215{,}171.60 - 214{,}038.53$	1,133.07
Within sum of squares	$\Sigma\Sigma(X^2) - \Sigma(\Sigma X)^2/n$, or $216{,}910 - 215{,}171.6$	1,738.40
Total sum of squares	$\Sigma\Sigma(X^2) - (\Sigma\Sigma X)^2/N$, or $216{,}910 - 214{,}038.53$	2,871.47

Second, we need to compute the mean sum of squares, which is simply an average sum of squares. These are the variance estimates that we need to eventually compute the all-important F ratio.

We do that by dividing each sum of squares by the appropriate number of degrees of freedom (df). Remember, degrees of freedom are an approximation of the sample or group size. We need two sets of degrees of freedom for ANOVA. For the between-group estimate, it is $k - 1$, where k equals the number of groups (in this case, there are 3 groups and 2 degrees of freedom), and for the within-group estimate, we need $N - k$, where N equals the total sample size (which means that the number of degrees of freedom is $30 - 3$, or 27). And the F ratio is simply a ratio of the mean sums of squares due to between-group differences over the mean sums of squares due to within-group differences, or $566.54/64.39 = 8.799$. This is the obtained F value.

Here's a summary table of the variance estimates used to compute the F ratio and how most F tables appear in professional journals and manuscripts.

Source	Sums of Squares	df	Mean Sums of Squares	F
Between groups	1,133.07	2	566.54	8.799
Within groups	1,738.40	27	64.39	
Total	2,871.47	29		

All that trouble for one little *F* ratio. But as we have said earlier, it's essential to do these procedures at least once by hand. It gives you the important appreciation of where the numbers come from and some insight into what they mean.

> Because you already know about *t* tests, you might be wondering how a *t* value (which is always used for the test between the difference of the means for two groups) and an *F* value (which is always more than two groups) might be related. Interestingly enough, an *F* value for two groups is equal to a *t* value for two groups squared,
>
> $$t^2 = F \text{ or } t = \sqrt{F}.$$
>
> Handy trivia question, right? But also useful if you know one and need to know the other.

5. Determination of the value needed for rejection of the null hypothesis using the appropriate table of critical values for the particular statistic.

As we have done before, we have to compare the obtained and critical values. We now need to turn to the table that lists the critical values for the *F* test, Table B.3 in Appendix B. Our first task is to determine the degrees of freedom for the numerator, which is $k - 1$, or $3 - 1 = 2$. Then, determine the degrees of freedom for the denominator, which is $N - k$, or $30 - 3 = 27$. Together, they are represented as $F_{(2, 27)}$.

The obtained value is 8.80, or $F_{(2, 27)} = 8.80$. The critical value at the .05 level with 2 degrees of freedom in the numerator (represented by columns in Table B.3) and 27 degrees of freedom in the denominator (represented by rows in Table B.3) is 3.36. So, at the .05 level, with 2 and 27 degrees of freedom for an omnibus test between the means of the three groups, the value needed for rejection of the null hypothesis is 3.36.

6. A comparison of the obtained value and the critical value is made.

The obtained value is 8.80, and the critical value for rejection of the null hypothesis at the .05 level that the three groups are different from one another (without concern for where the difference lies) is 3.36.

7. and 8. Decision time.

Now comes our decision. If the obtained value is more extreme than the critical value, the null hypothesis cannot be accepted. If the obtained value does not exceed the critical value, the null hypothesis is the most attractive explanation. In this case, the obtained value does exceed the critical value—it is extreme enough for us to say that the difference between the three groups is not due to chance. And if we did our experiment correctly, then what could the factor be that affected the outcome? Easy—the number of hours of preschool. We know the difference is due to a particular factor because the difference between the groups could not have occurred by chance, but instead is due to the treatment.

So How Do I Interpret $F_{(2, 27)} = 8.80$, $p < .05$?

- F represents the test statistic that was used.
- 2, 27 are the numbers of degrees of freedom for the between-group and within-group estimates.
- 8.80 is the obtained value using the formula we showed you earlier in the chapter.
- $p < .05$ (the really important part of this little phrase) indicates that the probability is less than 5% on any one test of the null hypothesis that the average scores of each group's language skills differ due to chance alone rather than the effect of the treatment. Because we defined .05 as our criterion for the research hypothesis being more attractive than the null hypothesis, our conclusion is that there is a significant difference among the three sets of scores.

TECH TALK (Really Important) Tech Talk

Imagine this scenario. You're a high-powered researcher at an advertising company, and you want to see if color makes a difference in sales. And you'll test this at the .05 level. So, you put together a brochure that is all black and white, one that is 25% color, the next 50%, then 75%, and, finally, 100% color, for five different levels. You do an ANOVA and find out that there is a difference. But because ANOVA is an omnibus test, you don't know where the source of the significant difference lies. So, you take two groups at time (such as 25% color and 75% color) and test them against each other. In fact, you test every combination of 2 against each other. Kosher? No way. This is called performing multiple *t* tests, and it is actually against the law in some jurisdictions. When you do this, the Type I error rate (which you set at .05) balloons depending on the number of tests you want to conduct. There are 10 possible comparisons (no color vs. 25%, no color vs. 50%, no color vs. 75%, etc.), and the real Type I error rate is equal to $1 - (1 - \alpha)^k$, where

α is the Type I error rate, which is .05 in this example, and

k is the number of comparisons.

So, instead of .05, the actual error rate that each comparison is being tested at is

$$1 - (1 - .05)^{10} = .40 \; (!!!!!).$$

Surely not .05. Quite a difference, no?

And Now . . . *Using Excel's FDIST and FTEST Functions*

Interestingly, just as with the TTEST, Excel does not have a function that computes the specific value of the test statistic. Also, **FDIST** and **FTEST** calculate values for only two groups of data (which is what we used the TTEST and TDIST functions and the ToolPak tools for in the previous chapter). So, we'll just move on to the ToolPak and the powerful ANOVA tools that it offers.

USING THE AMAZING ANALYSIS TOOLPAK TO COMPUTE THE *F* VALUE

ANOVA is more sophisticated than any of the other inferential tools that we have covered so far in *Statistics for People . . .* , and it's the ANOVA ToolPak tools that really shine and bring out the value of learning statistics through the use of Excel.

There are three different ANOVA options within the ToolPak, as follows:

- Anova: Single Factor,
- Anova: Two-Factor With Replication, and
- Anova: Two-Factor Without Replication.

We'll cover the first one in the following section and the other two in the next chapter when we cover the factorial analysis of variance. Hold on tight! We're using the data shown to you earlier in the chapter that also appear in Figure 12.3 as three columns, each column representing a different level of treatment (5, 10, and 20 hours per week).

Because some of you can't wait until that next chapter, the repeated part in the analysis of variance means that the same subjects are tested more than once—kind of like the dependent means in a *t* test.

	A	B	C
1	Group 1 Language Scores	Group 2 Language Scores	Group 3 Language Scores
2	87	87	89
3	86	85	91
4	76	99	96
5	56	85	87
6	78	79	89
7	98	81	90
8	77	82	89
9	66	78	96
10	75	85	96
11	67	91	93

Figure 12.3   Data for a Single Factor Analysis of Variance

1. Click Data → Data Analysis, and you will see the Data Analysis dialog box shown in Figure 12.4.

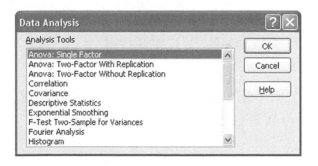

Figure 12.4   The Dialog Box That Gets us Started With the Analysis ToolPak

2. Click Anova: Single Factor and then click OK, and you will see the Anova: Single Factor statistics dialog box as shown in Figure 12.5.

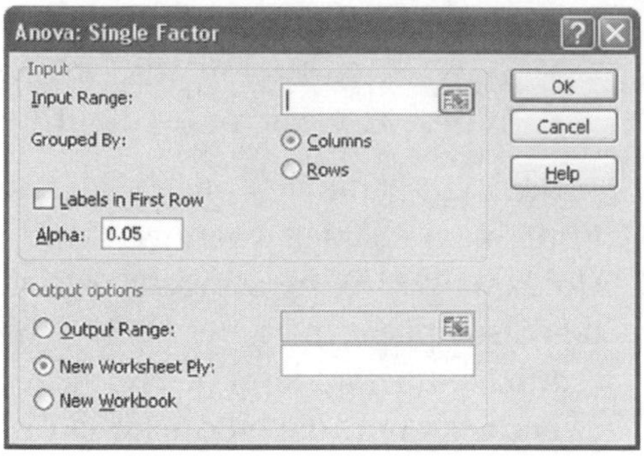

Figure 12.5   The Anova: Single Factor Dialog Box

3. In the Input Range, enter the cell addresses for the three groups of data. In the sample spreadsheet that you saw in Figure 12.3, the cell addresses are A1:C11 (and this includes the labels for each group with the label box checked).

4. Click whether the data are Grouped By: Columns or Rows (and it will almost always be grouped by columns as it is here).

5. Click the Labels box so that labels are included.

6. Click the Output Range button, and enter an address where you want the output located on the same worksheet as the data. In this example, we are placing the output beginning in Cell A13.

7. Click OK, and as you can see in Figure 12.6, you get a tidy summary of important data relating to this analysis, including the following output and what it means.

	A	B	C	D	E	F	G
1	Group 1 Language Scores	Group 2 Language Scores	Group 3 Language Scores				
2	87	87	89				
3	86	85	91				
4	76	99	96				
5	56	85	87				
6	78	79	89				
7	98	81	90				
8	77	82	89				
9	66	78	96				
10	75	85	96				
11	67	91	93				
12							
13	Anova: Single Factor						
14							
15	SUMMARY						
16	Groups	Count	Sum	Average	Variance		
17	Group 1 Language Scores	10	766	76.6	143.16		
18	Group 2 Language Scores	10	852	85.2	38.4		
19	Group 3 Language Scores	10	916	91.6	11.6		
20							
21							
22	ANOVA						
23	Source of Variation	SS	df	MS	F	P-value	F crit
24	Between Groups	1133.07	2.00	566.53	8.80	0.00	3.35
25	Within Groups	1738.40	27.00	64.39			
26							
27	Total	2871.47	29.00				

Figure 12.6   The Output for the Anova: Single Factor

Groups	The listing of each of the groups
Count	The number of observations in each group
Sum	The sum of the values
Average	The mean for each group
Variance	The variance for each group
Source of Variation	The source of error
SS	The sums of squares
df	Degrees of freedom associated with each source of error
MS	The mean square
F	The *F* value (the obtained value)
P-value	The level of significance of the *F* value
F crit	The value needed to reject the null hypothesis

As we showed you before, the *F* value is large enough that we would not expect it to be due to chance alone. So, our conclusion is that there is a difference in language development as a function of the number of hours spent in practice.

Summary

Analysis of variance (either single or factorial) is the most complex of all the inferential tests you will learn in *Statistics for People Who (Think They) Hate Statistics . . . Excel 2007 Edition*. It takes a good deal of concentration to perform the manual calculations, and even when you use Excel, you have to be on your toes to understand that this is an overall test, and one part will not give you information about differences between specific pairs of treatments. Only one more test between averages and that's a factorial ANOVA—the Holy Grail of ANOVAs—which can involve two or more factors, and here it comes in Chapter 13.

Time to Practice

1. Using the following table, provide three examples of a simple one-way ANOVA, two examples of a two-factor ANOVA, and one example of a three-factor ANOVA. We show you some examples. Be sure to identify the grouping and the test variable as we have done here.

Design	Grouping Variable(s)	Test Variable
Simple ANOVA	Four levels of hours of training—2, 4, 6, and 8 hours	Typing accuracy
	Enter Your Example Here	Enter Your Example Here
	Enter Your Example Here	Enter Your Example Here

(Continued)

(Continued)

Design	Grouping Variable(s)	Test Variable
Two-factor ANOVA	Two levels of training and gender (2 × 2 design)	Typing accuracy
	Enter Your Example Here	Enter Your Example Here
	Enter Your Example Here	Enter Your Example Here
Three-factor ANOVA	Two levels of training and two of gender and three of income	Voting attitudes
	Enter Your Example Here	Enter Your Example Here

2. Using the data in Chapter 12 Data Set 1 and Excel, compute the F ratio for a comparison between the three levels representing the average amount of time that swimmers practice weekly (<15 hours, 15–25 hours, and >25 hours) with the dependent or outcome variable being their time for the 100-yard freestyle. Answer the question whether practice time makes a difference. Either use the ToolPak or do it manually.

3. Stephen recognizes that there are different techniques for attracting attention to advertisements, and he wants to test three of these for the sample product: all color, all black and white, and a combination. Here are the data on the attractiveness of each product on a scale from 1 to 10. Now he wants to know if there is a difference between the three formats. Is there?

Color	B&W	Combination
10	4	9
8	5	8
7	4	8
8	3	9
9	3	8
6	4	7
7	5	8
6	6	9
6	5	9
7	7	10
8	6	10
7	5	9
6	4	8
5	5	9
6	4	10
7	4	10
7	3	8

Answers to Practice Questions

1.

Design	Grouping Variable(s)	Test Variable
Simple ANOVA	Four levels of hours of training—2, 4, 6, and 8 hours	Typing accuracy
	Three age groups—20-, 25-, and 30-year-olds	Strength
	Six levels of job types	Job performance
Two-factor ANOVA	Two levels of training and gender (2 × 2 design)	Typing accuracy
	Three levels of age (5, 10, and 15 years) and number of siblings	Social skills
Three-factor ANOVA	Curriculum type (Type 1 or Type 2), GPA (above or below 3.0), and activity participation (participates or not)	ACT scores

2. As you can see in Figure 12.7, an abbreviated output from the ToolPak, the means for the three groups are 58.05 seconds, 57.96 seconds, and 59.03 seconds, and the probability of this F value $[F_{(2, 33)} = .160]$ occurring by chance is .85, far above what we would expect due to the treatment. Our conclusion? The number of hours of practice makes no difference in how fast you swim!

	A	B	C	D	E	F	G
1	Anova: Single Factor						
2							
3	SUMMARY						
4	Groups	Count	Sum	Average	Variance		
5	<15 Hours Practice	10	580.5	58.05	29.88		
6	15-25 Hours Practice	13	753.5	57.96	22.68		
7	More than 25 Hours Practice	13	767.4	59.03	31.08		
8							
9							
10	ANOVA						
11	Source of Variation	SS	df	MS	F	P-value	F crit
12	Between Groups	8.87	2	4.43	0.16	0.85	3.28
13	Within Groups	914.06	33	27.70			
14							
15	Total	922.93	35				

Figure 12.7 Abbreviated Output From the ToolPak

3. There certainly is a difference. The results of the analysis are $F_{(2, 48)} = 63.36$, $p < .000$, which means that the likelihood of there being a difference due to anything other than the format is very, very low. And, if you look at the overall means of the three groups as follows:

Group	Average
Color	7.06
B&W	4.53
Combination	8.76

It's clear that the Combination format is the highest.

13 Two Too Many Factors

Factorial Analysis of Variance: A Brief Introduction

Difficulty Scale ☺ (some challenging ideas—but we're only touching on the main concepts here)

How much Excel? [excel] [excel] (some)

What you'll learn about in this chapter

- When to use analysis of variance with more than one factor
- All about main and interaction effects
- Using the Amazing Analysis ToolPak to perform a factorial analysis of variance

INTRODUCTION TO FACTORIAL ANALYSIS OF VARIANCE

How people make decisions has been one of the processes that has fascinated psychologists for decades. The data that have resulted from the studies have been applied to such broad fields as advertising, business, planning, and even theology. Miltiadis Proios and George Doganis investigated how the experience of being actively involved in the decision-making process (in a variety of settings) and age can have an impact on moral reasoning. The sample consisted of a total of 148 referees—56 who referee soccer, 55 who referee basketball, and 37 who referee handball. Their ages ranged from 17 to 50 years, and gender was not considered an important variable. Within the entire sample, about 8% had not had any experience in social, political, or athletic settings where they fully participated in the decision-making process; about 53% were active but did not fully participate; and about 39% were both active and did participate in the decisions made within that organization. A two-way (multivariate—see Chapter 17 for more about this) analysis of variance showed an interaction between experience and age on moral reasoning and goal orientation of referees.

Why a two-way analysis of variance? Easy—there were two independent factors, with the first being level of experience and the second being age. Here, just as with any analysis of variance procedure, there is

1. a test of the main effect for age,

2. a test of the main effect for experience, and

3. a test for the interaction between experience and age (which turned out to be significant).

The very cool thing about analysis of variance when more than one factor or independent variable is tested is that the researcher can look at the individual effects of each factor, but also the simultaneous effects of both, through what is called an interaction, which we talk about more later in this chapter.

Want to know more? Proios, M., & Doganis, G. (2003). Experiences from active membership and participation in decision-making processes and age in moral reasoning and goal orientation of referees. *Perceptual and Motor Skills, 96*(1), 113–126.

Two Flavors of Factorial ANOVA

There are two types of factorial ANOVAs. Excel calls one "Anova: Two-Factor With Replication" and the other "Anova: Two-Factor Without Replication." Both involve two factors, but the difference is in how many times one factor is tested across the same individuals. For example, in an analysis of variance *without replication,* we could test the effects of two factors: location of residence (urban or rural) and voting preference (Green party or not Green party). As an outcome, we'll use attitude toward environmental waste.

In this design (shown below), there are separate and independent observations in each of the four cells. No one can have his or her primary residence in both an urban and a rural area, nor can he or she belong to both a Green and a non-Green voting bloc. There's no replication.

		Primary Residence	
		Urban	*Rural*
Political Affiliation	Green		
	Not Green		

On the other hand, in an analysis of variance *with replication,* we could test the effects of two factors such as change over time (from September to July) and subject matter (math or spelling). As an outcome, we'll use achievement.

In this example, the replication is over the variable named time, because the same (here's where the replication comes in) subjects are being tested twice. And, here's the design.

		Time	
		September	*June*
Achievement	Math		
	Spelling		

More Excel

Guess what? Excel is about the coolest personal computer application ever invented. But it, too, like others, has its shortcomings. One is that the ANOVA tools, the ones with and without replication, don't provide equally clear output. So, rather than have to go into excruciating detail (and you are using *this* book to avoid such), we're only going to deal with the replication option because the option without replication does not provide a clear example of how this technique works. Take a look at our software sampler in Chapter 18 to find out about other programs that you might want to use.

The Path to Wisdom and Knowledge

Here's how you can use the flow chart shown in Figure 13.1 to select ANOVA (but this time with more than one factor) as the appropriate test statistic. Follow the highlighted sequence of steps.

As in Chapter 12, we have already decided that ANOVA is the correct procedure (examining differences in more than two levels of the independent variable), but because we have more than one factor, factorial ANOVA is the right choice.

More Excel

ANOVA with replication is also referred to as repeated measures ANOVA or within measures ANOVA, because the measure is repeated across people.

1. We are testing for differences between scores of the same participants.

2. The participants are being tested more than once.

3. We are dealing with two or more groups.

4. We are dealing with more than one factor or independent variable.

5. The appropriate test statistic is factorial analysis of variance, with replication or repeated across conditions.

A New Flavor of ANOVA

You know that ANOVA comes in at least one flavor, the simple analysis of variance we discussed in Chapter 12. There is one factor or one treatment variable (such as group membership) being explored, and there are more than two groups or levels within this factor or treatment variable.

Now, we bump up the entire technique a notch to include the exploration of more than one factor simultaneously and call this a **factorial analysis of variance.**

Let's look at a simple example that includes two factors, gender (male or female) and treatment (high- or low-impact exercises) and the outcome—a weight loss score. The treatment is a weight loss program that has two levels of involvement—high impact and low impact. And the same people experience both the high- and the low-impact conditions in a counterbalanced order (one half get low first, then high, and the other half, vice versa). Here's what the experimental design would look like. And keep in mind (again) that this analysis is with replication—where the impact treatment is the repeated measure.

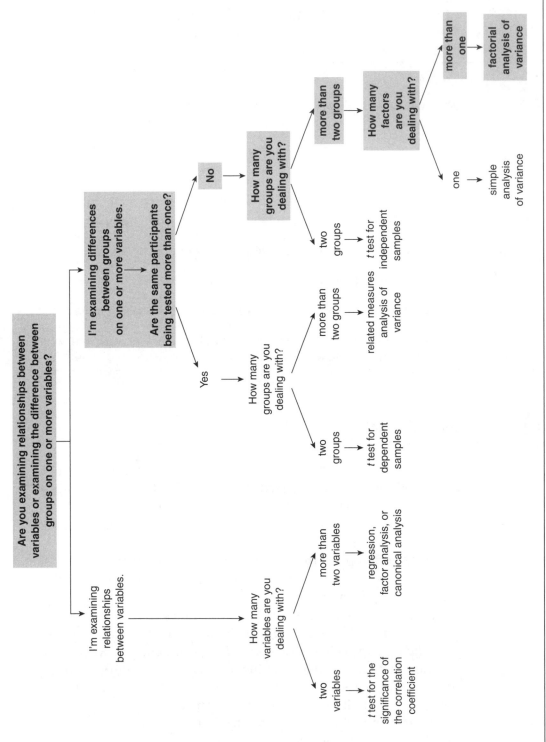

Figure 13.1 Determining That Factorial Analysis of Variance Is the Correct Test Statistic

TECH TALK

ANOVA designs that contain a factor where there is repetition and a factor where there is not are sometimes called mixed designs.

		Impact	
		High	Low
Gender	Male		
	Female		

Then, we will look at what main effects and an interaction look like. Not a lot of data analysis here until a bit later on in the chapter—mostly just look and learn.

There are three questions that you can ask and answer from this type of analysis:

1. Is there a difference between the levels of impact? Remember that each person participates at both the high and low level.

2. Is there a difference between the two levels of gender, male and female?

3. What is the effect of different levels of impact for males or females (and this is the famous interaction effect you will learn about shortly)?

Questions 1 and 2 deal with the presence of main effects, whereas Question 3 deals with the interaction between the two factors.

THE MAIN EVENT: MAIN EFFECTS IN FACTORIAL ANOVA

You might remember that the primary task of analysis of variance is to test for the difference between two or more groups. When an analysis of the data reveals a difference between the levels of any factor, we talk about there being a **main effect.** Here's an example where there are 10 participants in each of the four groups in the above example, for a total of 40. And here's what the results of the analysis look like. This is called a **source table.**

Source	Sum of Squares	df	Mean Square	F	Sig.
Impact	429.025	1	429.025	3.011	**.091**
Gender	3222.025	1	3222.025	22.612	**.000**
Impact × Gender	27.225	1	27.225	.191	**.665**
Error	5129.700	36	142.492		

Pay attention to only the *Source* and the *Sig.* columns (which are in boldface). The conclusion we can reach is that there is a main effect for gender ($p = .000$), no main effect for impact ($p = .091$), and no interaction between the two main factors ($p = .665$). So, as far as weight loss, it didn't matter whether one was in the high- or low-impact group, but it did matter if one were male or female. And because there was no interaction between the treatment factor and gender, there were no differential effects for treatment across gender. If you plotted the means of these values, you would get something that looks like this:

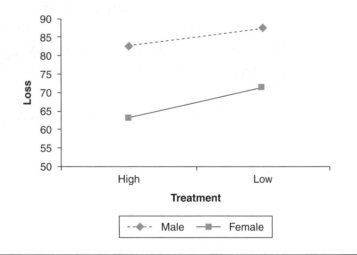

Figure 13.2   Mean Scores Across Treatments for Males and Females

You can see a big difference in distance on the loss axis between males and females (mean score for all males is 85.25 and for females, 67.30), but for treatment (if you computed the means), you would find there to be little difference (with the mean score across all highs being 73.00 and across all lows being 79.55). Now, of course, this is an analysis of variance, and the variability in the groups does matter, but in this example, you can see the differences between groups (such as males and females) within each factor (such as gender) and how they are reflected by the results of the analysis.

EVEN MORE INTERESTING: INTERACTION EFFECTS

OK—now let's move to the interaction. Let's look at a new source table that indicates men and women are affected differentially across treatments, indicating the presence of an **interaction effect.** And, indeed, you will see some very cool outcomes.

Source	Sums of Squares	df	Mean Square	F	Significance
Treatment	265.225	1	265.225	2.444	.127
Gender	207.025	1	207.025	1.908	.176
Treatment × Gender	1050.625	1	1050.625	9.683	.004
Error	3906.100	36	108.503		
Total	224321.000	39			

Here, there is no main effect for treatment or gender ($p = .127$ and .176, respectively), but yikes, there is one for the interaction ($p = .004$), which makes this a very interesting outcome. In effect, it does not matter if you are in the high- or low-impact treatment group, or if you are male or female, but it does matter if you consider both conditions simultaneously such that the treatment does have an impact differentially on the weight loss of males than on females.

Here's what a chart of the mean for each of the four groups looks like.

And, here's (Figure 13.3) what the actual means themselves look like (all compliments of the Excel AVERAGE function).

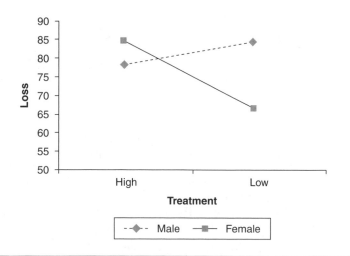

Figure 13.3   Averages Across Treatments for Males and Females

	Male	Female
High Impact	73.70	79.40
Low Impact	78.80	64.00

What to make of this? Well, the interpretation here is pretty straightforward. Here's what we can say, and being as smart as you are, you can recognize that these are the answers to the three questions we listed earlier.

- There is no main effect for type of exercise.
- There is no main effect for gender.
- There is a clear interaction between treatment and gender, which means females lose more weight than males under the high-impact treatment condition, and males lose more weight than females under the low-impact condition.

This is all pretty remarkable stuff. If you didn't know any better (and never read this chapter), you would think that all you have to do is a simple *t* test between the averages for males and females, and then another simple *t* test for the averages between those who participated in the high-impact and those who participated in the low-impact treatment—and you would have found nothing. But using the idea of an interaction between main factors, you find out that there is a differential effect—an outcome that would have gone unnoticed otherwise. Indeed, if you can bear the admission, interactions really are the most interesting outcomes in any factorial analysis of variance.

COMPUTING THE ANOVA F STATISTIC USING THE AMAZING ANALYSIS TOOLPAK

Here's a change for you. Throughout *Statistics for People Who (Think They) Hate Statistics . . . Excel 2007 Edition*, we have provided you with examples of how to perform particular techniques the old-fashioned way (by hand using a calculator) as well as with tools such as Excel. With the introduction of factorial ANOVA, we are illustrating the analysis using only Excel—nothing manual—not that it is any more of an intellectual challenge to complete a factorial ANOVA using a calculator, but it certainly is more laborious. Here are the data we'll use for this example and eventually show you how to use the ToolPak factorial or two-way ANOVA tool.

High-Impact Male	High-Impact Female	Low-Impact Male	Low-Impact Female
76	65	88	65
78	90	76	67
76	65	76	67
76	90	76	87
76	65	56	78
74	90	76	56
74	90	76	54
76	79	98	56
76	70	88	54
55	90	78	56

Before we get into using the ToolPak, first a statement of the null and research hypotheses.

There are actually three null hypotheses, shown here (Formulas 13.1a, 13.1b, and 13.1c) that state that there is no difference between the means for the two factors, and no interaction. Here we go.

First for the treatment:

$$H_0 : \mu_{high} = \mu_{low}, \tag{13.1a}$$

and now for gender:

$$H_0 : \mu_{male} = \mu_{female}, \tag{13.1b}$$

and now for the interaction between treatment and gender:

$$H_0 : \mu_{high \bullet male} = \mu_{high \bullet female} = \mu_{low \bullet female} = \mu_{low \bullet male}. \tag{13.1c}$$

The research hypotheses, shown in Formulas 13.2a, 13.2b, and 13.2c, state that there is a difference between the means of the groups, and there is an interaction. Here they are.

First for the treatment:

$$H_1 : \mu_{high} \neq \mu_{low}, \tag{13.2a}$$

and now for gender:

$$H_1 : \mu_{male} \neq \mu_{female}, \tag{13.2b}$$

and now for the interaction between treatment and gender:

$$H_1 : \mu_{high \bullet male} \neq \mu_{high \bullet female} \neq \mu_{low \bullet female} \neq \mu_{low \bullet male}. \tag{13.2c}$$

We'll use the Anova: Two-Factor With Replication option, and here are the steps. We'll use the above data, which are available on the Web site as Chapter 13 Data Set 1 (contained in Appendix C) and shown in Figure 13.4. If you want to actually follow along in using the ToolPak, then be sure that this data file is open, or enter the actual data and save the file.

	A	B	C
1		High	Low
2	Male	76	88
3		78	76
4		76	76
5		76	76
6		76	56
7		74	76
8		74	76
9		76	98
10		76	88
11		55	78
12	Female	65	65
13		90	67
14		65	67
15		90	87
16		65	78
17		90	56
18		90	54
19		79	56
20		70	54
21		90	56
22			

Figure 13.4   Chapter 13 Data Set 1 for the ANOVA Example

1. Click Data → Data Analysis and you will see the Data Analysis dialog box. Need a brush up on how to use the ToolPak? See "Tooling Around With the Amazing Analysis ToolPak" in Little Chapter 1b.

2. Click Anova: Two-Factor With Replication and then click OK, and you will see the Anova: Two-Factor With Replication dialog box as shown in Figure 13.5.

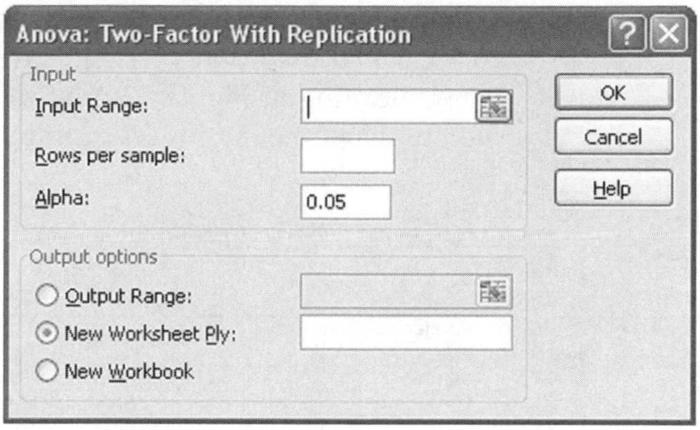

Figure 13.5 The Anova: Two-Factor With Replication Dialog Box

3. In the Input Range box, enter the range of data you want Excel to use in the computation of the correlations. As you can see in Figure 13.4, the data we want to analyze in Cells A1 through C21 and the column and row headings are included.

4. Enter the number of rows per sample. This is the same as the number of observations that you have in each cell, which is 10.

5. Enter the level of significance at which you would like the *F* value tested. In this example, we are going to use the .05 level.

6. Now click the Output Range button in the Output options section of the dialog box and enter the location where you want Excel to return the results of the analysis. In this example, we checked E1. The completed dialog box should appear as shown in Figure 13.6.

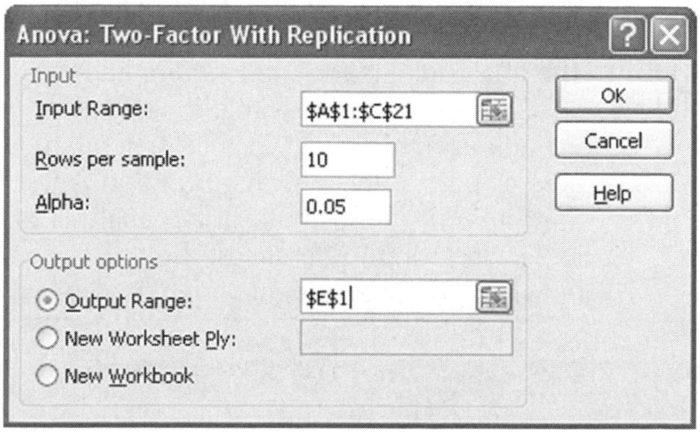

Figure 13.6 The Completed ANOVA Dialog Box

7. Click OK, and there it is, folks, the output you see in Figure 13.7. Truly a time to rejoice. (P.S. We did a bit of editing and house cleaning in Figure 13.7 by rounding cell contents and such to make the output easier to understand.)

	A	B	C	D	E	F	G	H	I	J	K
1		High	Low		Anova: Two-Factor With Replication						
2	Male	76	88								
3		78	76		SUMMARY	High	Low	Total			
4		76	76		*Male*						
5		76	76		Count	10	10	20			
6		76	56		Sum	737	788	1525			
7		74	76		Average	73.7	78.8	76.25			
8		74	76		Variance	44.46	121.96	85.67			
9		76	98								
10		76	88		*Female*						
11		55	78		Count	10	10	20			
12	Female	65	65		Sum	794	640	1434			
13		90	67		Average	79.4	64	71.7			
14		65	67		Variance	141.38	126.22	189.17			
15		90	87								
16		65	78		*Total*						
17		90	56		Count	20	20				
18		90	54		Sum	1531	1428				
19		79	56		Average	76.55	71.4				
20		70	54		Variance	96.58	175.2				
21		90	56								
22											
23					ANOVA						
24					*Source of Variation*	*SS*	*df*	*MS*	*F*	*P-value*	*F crit*
25					Sample	207.03	1	207.03	1.91	0.18	4.11
26					Columns	265.22	1	265.22	2.44	0.13	4.11
27					Interaction	1050.63	1	1050.63	9.68	0.00	4.11
28					Within	3906.10	36	108.50			
29											
30					Total	5428.98	39				
31											

Figure 13.7 A Completed Two-Way ANOVA Using Excel's Analysis ToolPak

Excel does not label the sources of variance ☹, but instead uses general terms. So, in this example, "Sample" represents gender, "Columns" represents level of impact, and "Interaction" represents the interaction between level of impact and gender.

TECH TALK The analysis that you are learning here is a univariate analysis of variance. This is an analysis that looks at only one dependent or outcome variable—in this case, weight loss score. If we had more than one variable as part of the research question (such as attitude toward eating), then it would be a multivariate analysis of variance, which not only looks at group differences, but also controls for the relationship between the independent variables. More about this in Chapter 18.

Summary

Now that we are done, done, done with testing differences between means, we'll move on to examine the significance of correlations, or the relationship between two variables.

Time to Practice

1. When would you use a factorial ANOVA rather than a simple ANOVA to test the significance of the difference between the average of two or more groups?

2. Create a 2 × 3 experimental design that would lend itself to a factorial ANOVA.

3. Using Excel, and using the data in Chapter 13 Data Set 2, complete the analysis and interpret the results. It is a 2 (2 levels of severity where Level 1 is severe and Level 2 is mild) × 3 (three levels of treatment where Level 1 is Drug #1, Level 2 is Drug #2, and Level 3 is Placebo) experiment. This is an ANOVA with replication because each participant received all three treatments, which is represented by the columns of data, and severity by the rows (or Sample as the Excel data analysis likes to call it).

Answers to Practice Questions

1. Easy. Factorial ANOVA is used only when you have more than one factor or independent variable! And actually, not so easy an answer to get (but if you get it, you really understand the material) when you hypothesize an interaction.

2. Here's one of many different possible examples. There are three levels of one treatment (or factor) and two levels of severity of illness. As far as our interpretation, in this data set, there is no main effect for severity, there is a main effect for treatment, and there is no interaction between the two main factors.

		Treatment		
		Drug #1	Drug #2	Placebo
Severity of Illness	Severe			
	Mild			

3. And the edited source table looks like this:

	ANOVA				
Source of Variation	SS	df	MS	F	p-value
Sample	0.075	1	0.075	0.037	0.848
Column	263.517	2	131.758	64.785	0.000
Interaction	3.150	2	1.575	0.774	0.463
Within	231.850	114	2.034		
Total	498.592	119			

And our conclusions? No main effect for severity, but a main effect for treatment and no interaction.

14

Cousins or Just Good Friends?

Testing Relationships Using the Correlation Coefficient

Difficulty Scale ☺☺☺☺ (easy—you don't even have to figure anything out!)

How much Excel? 📊 📊 (some)

- How to test the significance of the correlation coefficient
- The interpretation of the correlation coefficient
- Using the PEARSON function
- The important distinction between significance and meaningfulness (again!)

INTRODUCTION TO TESTING THE CORRELATION COEFFICIENT

In his research article on the relationship between the quality of a marriage and the quality of the relationship between the parent and the child, Daniel Shek tells us that there are at least two possibilities. First, a poor marriage might enhance parent–child relationships. This is because parents who are dissatisfied with their marriage might substitute their relationship with their children for emotional gratification. Or, according to the spillover hypothesis, a poor marriage might damage the parent–child relationship. This is because a poor marriage might set the stage for increased difficulty in parenting children.

Shek examined the link between marital quality and parent–child relationships in 378 Chinese married couples over a 2-year period. He found that higher levels of marital quality were related to higher levels of parent–child relationships; this was found for concurrent measures (at the present time) as well as longitudinal measures (over time). He also found that the strength of the relationship between parents and children was the same for both mothers and fathers. This is an obvious example of how the use of the correlation coefficient gives us the information we need about whether sets of variables are related to one another. Shek computed a whole bunch of different correlations across mothers and fathers as well at Time 1 and Time 2, but all with the same purpose: to see if there was a significant correlation between the variables. Remember that this does not say anything about the causal nature of the relationship, only that the variables are associated with one another.

Want to know more? Check out Shek, D. T. L. (1998). Linkage between marital quality and parent-child relationship. *Journal of Family Issues, 19,* 687–704.

The Path to Wisdom and Knowledge

Here's how you can use the flow chart to select the appropriate test statistic, the test for the correlation coefficient. Follow along the highlighted sequence of steps in Figure 14.1.

1. The relationship between variables, and not the difference between groups, is being examined.

2. Only two variables are being used.

3. The appropriate test statistic to use is the *t* test for the correlation coefficient.

COMPUTING THE TEST STATISTIC

Here's something you'll probably be pleased to read: The correlation coefficient can act as its own test statistic. This makes things much easier because you don't have to compute any test statistics, and examining the significance is very easy indeed. Let's use, as an example, the following data that examine the relationship between two variables, the quality of marriage and the quality of parent–child relationships.

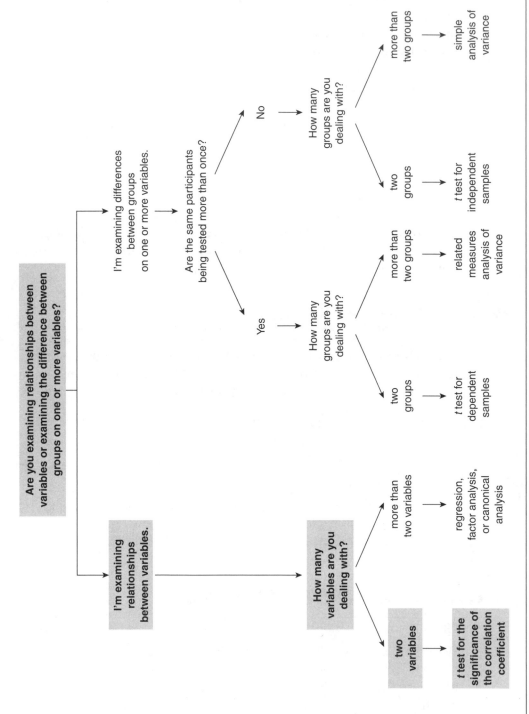

Figure 14.1 Determining That a *t* Test for the Correlation Coefficient Is the Correct Test Statistic

Quality of Marriage	Quality of Parent–Child Relationship
76	43
81	33
78	23
76	34
76	31
78	51
76	56
78	43
98	44
88	45
76	32
66	33
44	28
67	39
65	31
59	38
87	21
77	27
79	43
85	46
68	41
76	41
77	48
98	56
99	55
98	45
87	68
67	54
78	33

You can use Formula 5.1 from Chapter 5 (on page 117) to compute the Pearson correlation coefficient. When you do, you will find that $r = .393$. Now let's go through the steps of actually testing the value for significance and making a decision as to what the value means.

More Excel

You already know about the use of the CORREL function, as well as the Correlation tool in the Analysis ToolPak, and how easy they are to use to compute the correlation coefficient. Excel offers one other correlation tool that you should know about and that's the PEARSON function, which calculates the correlation coefficient much like the CORREL function and the Correlation ToolPak tool. For your purposes, all three of these compute the same value and are interchangeable. When you take Stat II, you may want to begin worrying about how CORREL and PEARSON are different (and the answer is in no significant way).

Here are the famous eight steps and the computation of the *t*-test statistic.

1. A statement of the null and research hypotheses.

The null hypothesis states that there is no relationship between the quality of the marriage and the quality of the relationship between parents and children. The research hypothesis is a two-tailed, non-directional research hypothesis because it posits that there is a relationship between the two variables, but the direction is not important. Remember that correlations can be positive (direct) or negative (indirect), and the most important characteristic of a correlation coefficient is its absolute value or size and not its sign (positive or negative).

The null hypothesis is shown in Formula 14.1:

$$H_0 : \rho_{xy} = 0. \tag{14.1}$$

The Greek letter ρ, or rho, represents the population estimate of the correlation coefficient.

The research hypothesis (shown in Formula 14.2) states that there is a relationship between the two values, and that the relationship differs from a value of 0.

$$H_1 : r_{xy} \neq 0. \tag{14.2}$$

One tail or two? It's pretty easy to conceptualize what a one-tailed versus a two-tailed test is when it comes to differences between means (remember, we discussed this in Chapter 7). And it may even be easy for you to understand a two-tailed test of the correlation coefficient (where any difference from zero is what's tested). But what about a one-tailed test? It's really just as easy. A directional test of the research hypothesis that there is a relationship posits that relationship as being either direct (positive) or indirect (negative). So, if you think that there is a positive correlation between two variables, then the test is one-tailed. Similarly, if you hypothesize that there is a negative correlation between two variables, the test is one-tailed as well. It's only when you don't predict the direction of the relationship that the test is two-tailed. Got it?

2. Setting the level of risk (or the level of significance or Type I error) associated with the null hypothesis.

The level of risk or Type I error or level of significance is .05.

3. and 4. Selection of the appropriate test statistic.

Using the flow chart shown in Figure 14.1, we determined that the appropriate test is for the correlation coefficient. In this instance, we do not need to compute a test statistic because the sample r value (r_{xy} = .393) is, for our purposes, the test statistic.

5. Determination of the value needed for rejection of the null hypothesis using the appropriate table of critical values for the particular statistic.

Table B.4 in Appendix B lists the critical values for the correlation coefficient.

Our first task is to determine the degrees of freedom (df), which approximate the sample size. For this particular test statistic, the degrees of freedom are $n - 2$, or $29 - 2 = 27$, where n is equal to the number of pairs used to compute the correlation coefficient. These are the degrees of freedom only for this test statistic and not necessarily for any other.

Using this number (27), the level of risk you are willing to take (.05), and a two-tailed test (because there is no direction to the research hypothesis), the critical value is .381 (using df = 25 because it is more conservative [and closer]). So, at the .05 level, with 27 degrees of freedom for a two-tailed test, the value needed for rejection of the null hypothesis is .381.

TECH TALK

OK, we cheated a little. Actually, you can compute a t value (just like for the test for the difference between means) for the significance of the correlation coefficient. The formula is not any more difficult than any you have dealt with up to now, but you won't see it here. The point is that some smart statisticians have computed the critical r value for different sample sizes (and, likewise, degrees of freedom) for one- and two-tailed tests at different levels of risk (.01, .05), as you see in Table B.4. So, if you are reading along in your journal and see that a correlation was tested using a t value, you'll now know why.

---

6. A comparison of the obtained value and the critical value is made.

The obtained value is .393, and the critical value for rejection of the null hypothesis that the two variables are not related is .381.

7. and 8. Making a decision.

Now comes our decision. If the obtained value (or the value of the test statistic) is more extreme than the critical value (or the tabled value), the null hypothesis cannot be accepted. If the obtained value does not exceed the critical value, the null hypothesis is the most attractive explanation.

In this case, the obtained value (.393) does exceed the critical value (.381)—it is extreme enough for us to say that the relationship between the two variables (quality of marriage and quality of parent–child relationships) did occur by something other than chance.

So How Do I Interpret $r_{(27)}$ = .393, p < .05?

- r represents the test statistic that was used.
- 27 is the number of degrees of freedom.
- .393 is the obtained value using the formula we showed you in Chapter 5. You can also use the CORREL function, the PEARSON function, or the Correlation tool in the Analysis ToolPak.
- $p < .05$ (the really important part of this little phrase) indicates that the probability is less than 5% on any one test of the null hypothesis that the relationship between the two variables is due to chance alone. Because we defined .05 as our criterion for the research hypothesis being more attractive than the null hypothesis, our conclusion is that there is a significant relationship between the two variables. This means that as the level of marital quality increases, so does the level of quality of the parent–child relationship. Similarly, as the level of marital quality decreases, so does the level of quality of the parent–child relationship.

> Remember way back in Chapter 6 that we used correlations to assess the reliability and the validity of different types of measurement tools? Correlations are also used in a variety of more sophisticated statistical techniques that fall under the general category of data reduction (such as factor analysis). So, if you don't get it already, correlations (even if they can sometimes be overinterpreted—as we review below—remember the ice cream and crime thing) are very useful and often used.

Causes and Associations (Again!)

You'd have thought that you heard enough of this already, but this is so important that we really can't emphasize it enough. So, we'll emphasize it again. Just because two variables are related to one another (as in the above example), this has no bearing on whether one causes the other. In other words, having a terrific marriage of the highest quality in no way ensures that the parent–child relationship will be of a high quality as well. These two variables may be correlated because they share some traits that might make a person a good husband or wife and also a good parent (patience, understanding, willingness to sacrifice), but it's certainly possible to see how someone can be a good husband or wife and have a terrible relationship with his or her children.

Remember the crime and ice cream example from Chapter 5? It's the same here. Just because things are related and share something in common with one another has no bearing on whether there is a causal relationship between the two.

Significance Versus Meaningfulness (Again, Again!)

In Chapter 5, we reviewed the importance of the use of the coefficient of determination for understanding the meaningfulness of the correlation coefficient. You may remember that you square the correlation coefficient to determine the amount of variance accounted for by one variable in another variable. In Chapter 9, we also went over the general issue of significance versus meaningfulness.

But we should mention and discuss this topic again. Even if a correlation coefficient is significant (as was the case in the example in this chapter), it does not mean that the amount of variance accounted for is meaningful. For example, in this case, the coefficient of determination for a simple Pearson correlation value of .393 is equal to .154, indicating that 15.4% of the variance is accounted for and a whopping 84.6% of the variance is not. It leaves lots of room for doubt, doesn't it?

So, even though we know that there is a positive relationship between the quality of a marriage and the quality of a parent–child relationship and they tend to "go" together, the relatively small correlation of .393 indicates that there are lots of other things going on in that relationship that may be important as well. So, if ever you wanted to apply a popular saying to statistics, "What you see is not always what you get."

Summary

Correlations are powerful tools that point out the direction of a relationship and help us to better understand what two different outcomes share with one another. Remember that correlations work only when you are talking about associations and never when you are talking about causal effects.

Time to Practice

1. Given the following information, use Table B.4 in Appendix B to determine whether the correlations are significant and how you would interpret the results.

   a. The correlation between speed and strength for 20 women is .567. Test these results at the .01 level using a one-tailed test.

   b. The correlation between the number correct on a math test and the time it takes to complete the test is −.45. Test whether this correlation is significant for 80 children at the .05 level of significance. Choose either a one- or two-tailed test and justify your choice.

   c. The correlation between number of friends and grade point average (GPA) for 50 adolescents is .37. Is this significant at the .05 level for a two-tailed test?

2. Use the data in Chapter 14 Data Set 1 to answer the questions below. Do the analysis manually or using the ToolPak.

   a. Compute the correlation between motivation and GPA.

   b. Test for the significance of the correlation coefficient at the .05 level using a two-tailed test.

   c. True or false? The more highly you are motivated, the more you will study. Which did you select and why?

3. Discuss the general idea that just because two things are correlated, it does not mean that one causes the other. Provide an example (other than ice cream and crime!).

Answers to Practice Questions

1a. With 18 degrees of freedom ($df = n − 2$) at the .01 level, the critical value for rejection of the null hypothesis is .516. There is a significant correlation between speed and strength, and the correlation accounts for 32.15% of the variance.

1b. With 78 degrees of freedom at the .05 level, the critical value for rejection of the null hypothesis is .183 for a one-tailed test. There is a significant correlation between number correct and time. A one-tailed test was used because the research hypothesis was that the relationship was indirect or negative, and approximately 20% of the variance is accounted for.

1c. With 48 degrees of freedom at the .05 level, the critical value for rejection of the null hypothesis is .273 for a two-tailed test (we used Table B.4). There is a significant correlation between number of friends a child might have and GPA, and the correlation accounts for 13.69% of the variance.

2a. and b. We used the ToolPak to compute the correlation as .434, significant at the .017 level using a two-tailed test. Figure 14.2 shows the final output from the analysis.

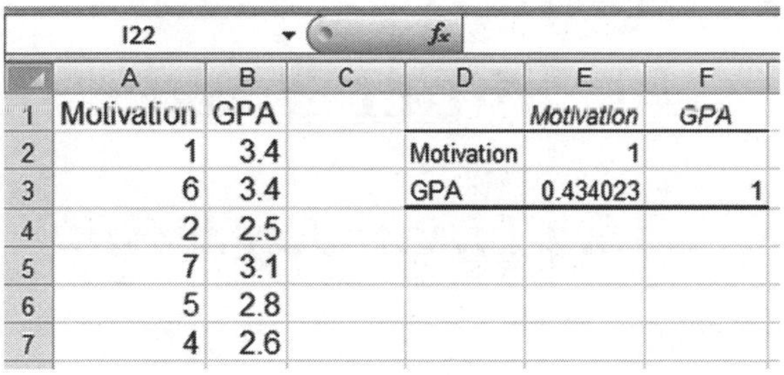

	A	B	C	D	E	F
					122 ▾ *fx*	
1	Motivation	GPA			*Motivation*	*GPA*
2	1	3.4		Motivation	1	
3	6	3.4		GPA	0.434023	1
4	2	2.5				
5	7	3.1				
6	5	2.8				
7	4	2.6				

Figure 14.2   Correlation ToolPak Output for Chapter 14 Data Set 1

2c. True. The more motivated you are, the more you will study, and the more you study, the more you are motivated. But (and this is a big "but") studying more does not cause you to be more highly motivated, nor does being more highly motivated *cause* (and that's the key word) you to study more.

3. The example here is the number of hours you study, and your performance on your first test in statistics. These variables are not causally related. For example, you will have classmates who studied for hours and did poorly because they never understood the material, and classmates who did very well without any studying at all because they had some of the same material in another class. Just imagine if we forced someone to stay at his or her desk and study for 10 hours each of four nights before the exam. Would that ensure that he or she would get a good grade? Of course not. Just because they are related does not mean that one causes the other.

15 Predicting Who'll Win the Super Bowl

Using Linear Regression

Difficulty Scale ☺ (as hard as they get!)

How much Excel? ⊠ ⊠ ⊠ ⊠ ⊠ (a ton)

What you'll learn about in this chapter

- How prediction works and how it can be used in the social and behavioral sciences
- How and why linear regression works when predicting one variable from another
- How to judge the accuracy of predictions
- Using the INTERCEPT and SLOPE functions
- What multiple regression is and why it is useful

WHAT IS PREDICTION ALL ABOUT?

Here's the scoop. Not only can you compute the degree to which two variables are related to one another (by computing a correlation coefficient as we did in Chapter 5), but you can also use these correlations as the basis for the prediction of the value of one variable from the value of another. This is a very special case of how correlations can be used, and it is a very powerful tool for social and behavioral sciences researchers.

The basic idea is to use a set of previously collected data (such as data on variables *X* and *Y*), calculate the degree to which these variables are correlated with one another, and then use that correlation and the knowledge of *X* to predict *Y*. Sound difficult? It's not really, especially once you see it illustrated.

For example, a researcher collects data on total high school grade point average (GPA) and first-year college GPA for 400 students in their freshman year at the state university. He computes the correlation between the two variables. Then, he uses the techniques you'll learn about later in this chapter to take a *different* set of high school GPAs and (knowing the relationship between high school GPA and first-year college GPA from the previous set of students) predict what first-year GPA should be for a new sample of 400 students. Pretty nifty, huh?

Here's another example. A group of teachers is interested in finding out how well retention works. That is, do children who are retained in kindergarten (and not passed on to first grade) eventually do better in first grade? Once again, these teachers know the correlation between being retained and first-grade performance; they can apply it to a new set of students and predict first-grade performance based on kindergarten performance. How does this work? Easy. Data are collected on past events (such as the existing relationship between two variables) and then applied to a future event given knowledge of only one variable. It's easier than you think.

> The higher the absolute value of the correlation coefficient, the more accurate the prediction is of one variable from the other based on that correlation, because the more two variables share in common, the more you know about the second variable from your knowledge of the first variable. And you may already surmise that when the correlation is perfect (+1.0 or –1.0), then the prediction is perfect as well. If $r_{xy} = -1.0$ or +1.0, and if you know the value of *X*, then you also know the exact value of *Y*. Likewise, if $r_{xy} = -1.0$ or +1.0, and you know the value of *Y*, then you also know the exact value of *X*. Either way works just fine.

What we'll do in this chapter is go through the process of using linear regression to predict a *Y* score from an *X* score. We'll begin by discussing the general logic that underlies prediction, then go to a review of some simple line-drawing skills, and, finally, discuss the prediction process using specific examples.

THE LOGIC OF PREDICTION

Before we begin with the actual calculations and show you how correlations are used for prediction, let's create the argument why and how prediction works. Then, we will continue with the example of predicting college GPA from high school GPA.

Prediction is an activity that computes future outcomes from present ones. When we want to predict one variable from another, we need to first compute the correlation between the two variables. Table 15.1 shows the data we will be using in this example. Figure 15.1 shows the scatterplot (see Chapter 5) of the two variables that are being computed.

TABLE 15.1 Total High School GPA and First-Year College GPA Are Correlated

High School GPA	First-Year College GPA
3.50	3.30
2.50	2.20
4.00	3.50
3.80	2.70
2.80	3.50
1.90	2.00
3.20	3.10
3.70	3.40
2.70	1.90
3.30	3.70

To predict college GPA from high school GPA, we have to create a **regression equation** and use that to plot what is called a **regression line**. A regression line reflects our best guess as to what score on the *Y* variable (college GPA) would be predicted by a score on the *X* variable (high school GPA). For all the data you see in Table 15.1, it's the line that minimizes the distance between the line and each of the points on the predicted (*Y*) variable. You'll learn shortly how to draw that line, shown in Figure 15.2. What does this regression line represent?

First, it's the regression of the *Y* variable on the *X* variable. In other words, *Y* (college GPA) is being predicted from *X* (high school GPA). This regression line is also called the **line of best fit.** The line best fits these data because it minimizes the distance between each individual point and the regression line. For example, if you take all of these points and try to find the line that best fits them all at once, the line you see in Figure 15.2 is the one you would use.

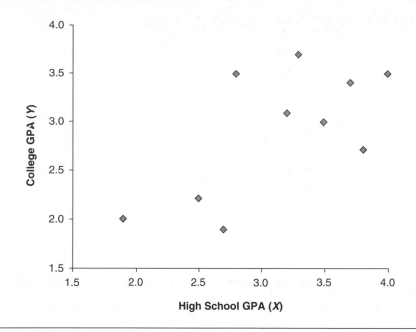

Figure 15.1   Scatterplot of High School GPA and College GPA

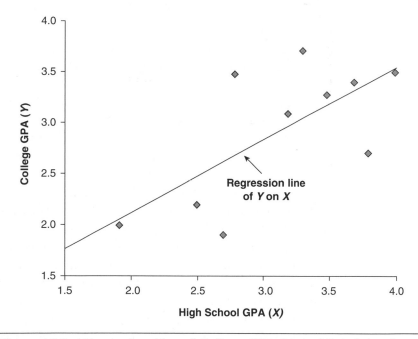

Figure 15.2   Regression Line of College GPA (*Y*) on High School
GPA (*X*)

Second, it's the line that allows us our best guess (at estimating what college GPA would be, given each high school GPA). For example, if high school GPA is 3.0, then college GPA should be around (remember, this is only an eyeball prediction) 2.8. Take a look at Figure 15.3 to see how we did this. We located the predictor value

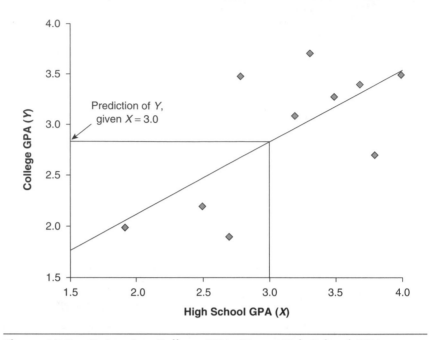

Figure 15.3   Estimating College GPA Given High School GPA

(3.0) on the *x*-axis, then drew a perpendicular line from the *x*-axis to the regression line, then drew a horizontal line to the *y*-axis and *estimated* what the value would be.

Third, the distance between each individual data point and the regression line is the **error in prediction**—a direct reflection of the correlation between the two variables. For example, if you look at data point 3.3, 3.7 (marked in Figure 15.4), you can see that this *X*, *Y* data point is above the regression line. The distance between that point and the line is the error in prediction, as marked in Figure 15.4, because if the prediction were perfect, then all the predicted points would fall where? Right on the regression or prediction line. You might think, at first glance, that the line should be drawn *perpendicular* to the regression line. Although that seems to make sense, it actually would not give you a true reading of the amount of error because the error can be computed based only on the distance from the data point to the *axis*, not the line.

Fourth, if the correlation were perfect (and the scale for the *x*- and *y*-axes were the same), all the data points would align themselves along a 45° angle, and the regression line would pass through each point (just like we said in the third point above).

Given the regression line, we can use it to predict any future score. That's what we'll do right now—create the line and then do some prediction work.

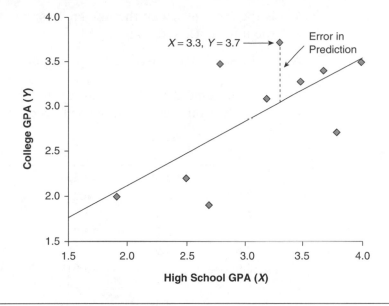

Figure 15.4 Prediction Is Rarely Perfect: Estimating the Error in Prediction

DRAWING THE WORLD'S BEST LINE (FOR YOUR DATA)

The simplest way to think of prediction is determining the score on one variable (which we'll call *Y*—the **criterion** or **dependent variable**) from the value of another score (which we'll call *X*—the **predictor** or **independent variable**).

The way that we find out how well *X* can predict *Y* is through the creation of the regression line we mentioned earlier in this chapter. This line is created from data that have already been collected. The equations are then used to predict scores using a new value for *X*, the predictor variable.

Formula 15.1 shows the general formula for the regression line, which may look familiar because you probably used it in your high school and college math courses. It's the same as the formula for any straight line:

$$Y' = bX + a, \tag{15.1}$$

where

Y' is the predicted score of *Y* based on a known value of *X*;

b is the slope, or direction, of the line;

a   is the point at which the line crosses the *y*-axis (also called the intercept); and

X   is the score being used as the predictor.

Let's use the same data shown earlier in Table 15.1 with a few more calculations thrown in that we will need.

	X	Y	X²	Y²	XY
	3.5	3.3	12.25	10.89	11.55
	2.5	2.2	6.25	4.84	5.50
	4.0	3.5	16.00	12.25	15.00
	3.8	2.7	15.44	7.29	10.26
	2.8	3.5	7.84	12.25	9.80
	1.9	2.0	3.61	4.00	3.80
	3.2	3.1	10.24	9.61	9.92
	3.7	3.4	13.69	11.56	12.58
	2.7	1.9	7.29	3.61	5.13
	3.3	3.7	10.89	13.69	12.21
Total	31.4	29.3	102.50	89.99	94.75

ΣX or the sum of all the X values, is 31.4.

ΣY or the sum of all the Y values, is 29.3.

ΣX^2 or the sum of each X value squared, is 102.5.

ΣY^2 or the sum of each Y value squared, is 89.99.

ΣXY or the sum of the products of X and Y, is 94.75.

Formula 15.2 is used to compute the slope of the regression line (*b* in the equation for a straight line):

$$b = \frac{\Sigma XY - (\Sigma X \Sigma Y / n)}{\Sigma X^2 - \left[(\Sigma X)^2 / n \right]}. \tag{15.2}$$

In Formula 15.3, you can see the computed value for *b*, the slope of the line.

$$b = \frac{94.74 - (31.4 \times 29.3/10)}{102.5 - \left[(31.4)^2 / 10 \right]}, \tag{15.3}$$

$$b = \frac{2.74}{3.904} = .702.$$

Formula 15.4 is used to compute the point at which the line crosses the *y*-axis (*a* in the equation for a straight line):

$$a = \frac{\Sigma Y - b\Sigma X}{n}. \tag{15.4}$$

In Formula 15.5, you can see the computed value for *a*, the intercept of the line.

$$a = \frac{29.3 - (.702 \times 31.4)}{10}, \tag{15.5}$$

$$a = \frac{7.26}{10} = .726.$$

Now, if we go back and substitute *b* and *a* into the equation for a straight line ($Y = bX + a$), we come up with the final regression line:

$$Y' = .702X + .726. \tag{15.6}$$

Why the Y' and not just a plain Y? Remember, we are using X to predict Y, and Y' (read: **Y prime**) is the predicted and not the actual value of Y. So, now that we have this equation, what can we do with it? Predict Y, what else?

For example, let's say that a current high school senior's GPA equals 2.8 (or $X = 2.8$). If we substitute the value of 2.8 into the equation, we get the following formula:

$$Y' = .702(2.8) + .726 = 2.69. \tag{15.7}$$

So, 2.69 is the predicted value of Y (or Y') given X is equal to 2.8. Now, for any X score, we can easily and quickly compute a predicted Y score.

More Excel

You can use the values in this formula and the predicted scores to compute predicted values. That's most of what we just talked about. But you can also plot this line to show how well the scores actually fit the data (what you are trying to predict) from which you are predicting. Take a look at Figure 15.2, which is a plot of the High School–College GPA data we showed you earlier and includes a Trend Line (which is another name for the regression line). How did we get this line? Easy. We used the same charting skills you learned in Chapter 4 to create a scatterplot; then we selected the Trend Line from the Chart option and selected linear. Poof! Done!

You can see the trend is positive (in that the line has a positive slope) and that the correlation is .6835—very positive. And you can see how the data points do not align directly on the line, but they are pretty close, which indicates that there is a relatively small amount of error.

TECH TALK

Not all lines that fit best between a bunch of data points are straight. Rather, they could be curvilinear, just like you can have a curvilinear relationship as we discussed in Chapter 5. For example, the relationship between anxiety and performance is such that when people are not at all anxious or very anxious, they don't perform very well. But if they're moderately anxious, then performance can be maximized. The relationship between these two variables is curvilinear, and the prediction of Y from X takes that into account.

More Excel

There are a lot of functions that deal with linear regression of some sort, such as **TREND, LINEST, FREQUENCY, STEYX,** and **FORECAST** (you can use Excel Help [press the F1 key] to find out what they do and how to use them). We're going to deal only with the two that best reflect what we have discussed throughout this chapter. The first is **SLOPE,** which computes the slope of the line or what was named b in the equation $Y' = bX + a$, and the second is **INTERCEPT,** which will compute a in the same equation or the point where the line crosses the x-axis.

And Now . . . Using Excel's SLOPE Function

The SLOPE function computes the slope of the line, or what we called b in the linear regression equation we discussed earlier in the chapter. To compute SLOPE, follow these steps.

1. We're using the data you see in Figure 15.5, which represents two sets of scores.

	A	B
1	High School GPA	First Year College GPA
2	3.5	3.3
3	2.5	2.2
4	4.0	3.5
5	3.8	2.7
6	2.8	3.5
7	1.9	2.0
8	3.2	3.1
9	3.7	3.4
10	2.7	1.9
11	3.3	3.7

Figure 15.5   Data for the Regression Example

2. Select the cell into which you want to enter the SLOPE function. In this example, we are going to place the value in Cell B13. We also placed a label in Cell A13 so we keep everything clear and understandable.

3. Click Formulas → More Functions → Statistical → SLOPE, and you will see the Function Arguments dialog box as shown in Figure 15.6.

Function Arguments [?][X]

SLOPE

Known_y's [   ] [▦] = array
Known_x's [   ] [▦] = array

=

Returns the slope of the linear regression line through the given data points.

Known_y's is an array or cell range of numeric dependent data points and can be numbers or names, arrays, or references that contain numbers.

Formula result =

Help on this function   [OK]   [Cancel]

Figure 15.6   The Function Arguments Dialog Box

4. Enter the range of cells for the Known_y's (the variable you want to predict), which is college GPA, or the Y variable, which is B2 through B11, as you see in Figure 15.5.

5. Enter the range of cells for the Known_x's (the variable from which you are predicting), which is high school GPA, or the X variable, which is A2 through A11, and the completed dialog box should appear as shown in Figure 15.7.

Figure 15.7　The Completed Dialog Box for the Excel SLOPE Function

5. Click OK, and you will see the slope of the best fitting regression line in Cell B13, as shown in Figure 15.8. Be sure to notice the entire function, as shown in the formula bar in Figure 15.8.

B13	fx	=SLOPE(B2:B11,A2:A11)

	A	B	C
1	High School GPA	First Year College GPA	
2	3.5	3.3	
3	2.5	2.2	
4	4.0	3.5	
5	3.8	2.7	
6	2.8	3.5	
7	1.9	2.0	
8	3.2	3.1	
9	3.7	3.4	
10	2.7	1.9	
11	3.3	3.7	
12			
13		0.703893443	

Figure 15.8　The Completed SLOPE Function

And Now . . . Using Excel's INTERCEPT Function

The INTERCEPT function computes the location where the regression line crosses the x-axis, or the *a* in the regression equation we discussed earlier in this chapter.

To compute INTERCEPT, follow these steps.

1. We're using the same data as we used to illustrate the calculation of SLOPE.

2. Select the cell into which you want to enter the INTERCEPT function. In this example, we are going to place the value in Cell B14.

We also placed a label in Cell A14 so we keep everything clear and understandable.

3. Click Formulas → More Functions → Statistical → INTERCEPT, and you will see the Function Arguments dialog box as shown in Figure 15.9.

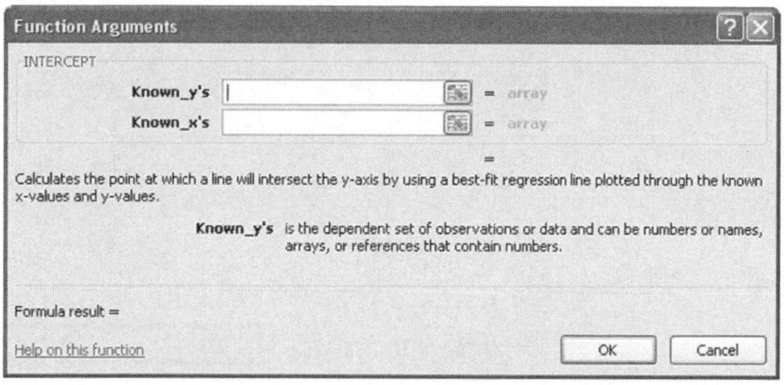

Figure 15.9   The Function Arguments Dialog Box

4. Enter the range of cells for the Known_y's (the variable you want to predict), which is college GPA, or the *Y* variable, which is B2 through B11, as you see in Figure 15.5.

5. Enter the range of cells for the Known_x's (the variable from which you are predicting), which is high school GPA, or the *X* variable, which is A2 through A11, and the completed dialog box should appear as shown in Figure 15.10.

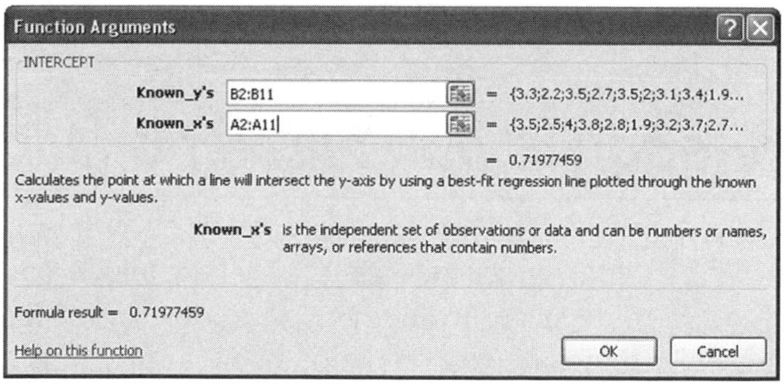

Figure 15.10   The Completed Dialog Box for INTERCEPT

6. Click OK, and you will see the intercept of the best fitting regression line in Cell B13, as shown in Figure 15.11. Be sure to notice the entire function, as shown in the formula bar in Figure 15.11.

B14	▾	fx	=INTERCEPT(B2:B11,A2:A11)

◢	A	B	C	
1	High School GPA	First Year College GPA		
2	3.5	3.3		
3	2.5	2.2		
4	4.0	3.5		
5	3.8	2.7		
6	2.8	3.5		
7	1.9	2.0		
8	3.2	3.1		
9	3.7	3.4		
10	2.7	1.9		
11	3.3	3.7		
12				
13	Slope (b)	0.703893443		
14	Intercept (a)	0.71977459		
15				

Figure 15.11   The Completed INTERCEPT Function

So, just as we manually calculated the values for the slope (b) and the intercept (a) of the regression line, we used SLOPE and INTERCEPT to do the same. Now, if we want to be especially cute (and efficient), we can construct a worksheet as you see in Figure 15.12, that in Cell B18 contains a formula that incorporates both functions to compute an estimated or predicted score from an actual X score. Be sure to look at the formula in the formula bar to see how we incorporated the calculations that Excel did for us. Magic!

HOW GOOD IS OUR PREDICTION?

How can we measure how good a job we have done predicting one outcome from another? We know that the higher the absolute magnitude of the correlation between two variables, the better the prediction. In theory, that's great. But being practical, we can also look at the difference between the predicted value (Y') and the actual value (Y) when we first compute the formula of the regression line.

For example, if the formula for the regression line is $Y' = .702X + .726$, the predicted Y (or Y') for an X value of 2.8 is $.702(2.8) + .726$, or 2.69. We know that the actual Y value that corresponds to an X value is 3.5 (from the data set shown in Table 15.1). The difference between 3.5 and 2.69 is .81 and is known as an **error of estimate**.

B18		▼	f_x	=B13*A18+B14
	A		B	
1	High School GPA		First Year College GPA	
2	3.5		3.3	
3	2.5		2.2	
4	4.0		3.5	
5	3.8		2.7	
6	2.8		3.5	
7	1.9		2.0	
8	3.2		3.1	
9	3.7		3.4	
10	2.7		1.9	
11	3.3		3.7	
12				
13	Slope (b)		0.703893443	
14	Intercept (a)		0.71977459	
15				
16				
17	Actual Scpre		Predicted Score	
18		3.25	3.007428279	

Figure 15.12 Predicting a Score Based on the INTERCEPT and SLOPE Functions

If we take all of these differences, we can compute a measure of how much each data point (on the average) differs from the predicted data point, or the **standard error of estimate.** It's like a standard deviation of all the error scores. This value tells us how much imprecision there is in our estimate. As you might expect, the higher the correlation between the two values (and the better the prediction), the lower this error will be. In fact, if the correlation between the two variables is perfect (either +1 or −1), then the standard error of estimate is 0. Why? Because prediction is perfect, all of the actual data points fall on the regression line, and there's no error in estimating Y from X.

The predicted Y', or dependent variable, need not always be a continuous one, such as height, test score, or problem-solving skills. It can be a categorical variable, such as admit/don't admit, Level A/Level B, or Social Class 1/Social Class 2. The score that's used in the prediction is "dummy coded" to be a 1 or a 2 and then used in the same equation.

THE MORE PREDICTORS THE BETTER? MAYBE

All of the examples that we have used so far in the chapter have been for one criterion or outcome measure and one predictor variable. There is also the case of regression where more than one predictor or independent variable is used to predict a particular outcome. If one variable can predict an outcome with some degree of accuracy, then why couldn't two do a better job?

For example, if high school GPA is a pretty good indicator of college GPA, then how about high school GPA plus number of hours of extracurricular activities? So, instead of

$$Y' = bX + a, \tag{15.8}$$

the model for the regression equation becomes

$$Y' = bX_1 + bX_2 + a, \tag{15.9}$$

where

Y' is the value of the predicted score,

X_1 is the value of the first independent variable,

X_2 is the value of the second independent variable, and

b is the regression weight for that particular variable.

As you may have guessed, this model is called **multiple regression**. So, in theory anyway, you are predicting an outcome from two independent variables, rather than one. But you want to add additional independent variables only under certain conditions. First, any variable you add has to make a unique contribution to understanding the dependent variable. Otherwise, why use it? What do we mean by unique? The additional variable needs to explain differences in the predicted variable that the first predictor does not. That is, the two variables in combination would have to predict Y better than any one of the variables would do alone.

In our example, level of participation in extracurricular activities could make a unique contribution. But should we add a variable such as the number of hours each student studied in high school as a third independent variable or predictor? Because number of hours of study is probably highly related to high school GPA (another of our predictor variables, remember?), study time probably would not add very much to the overall prediction of college GPA. We might be better off looking for another variable (such as ratings on letters of recommendation) rather than spending our time collecting the data on study time.

The Big Rule When It Comes to
Using Multiple Predictor Variables

If you are going to use more than one predictor variable, try to keep the following two important guidelines in mind:

1. When selecting an independent variable to predict an outcome, select a predictor variable (X) that is related to the predicted variable (Y). That way, the two share something in common (remember, they should be correlated).

2. When selecting more than one independent variable (such as X_1 and X_2), try to select variables that are independent or uncorrelated with one another but are both related to the outcome or predicted (Y) variable.

In effect, you want only independent or predictor variables that are related to the dependent variable and are unrelated to each other. That way, each one makes as unique a contribution as possible in predicting the dependent or predicted variable.

TECH TALK How many predictor variables are too many? Well, if one variable predicts some outcome, and two is even more accurate, then why not three, four, or five predictor variables? In practical terms, every time you add a variable, an expense is incurred. Someone has to go collect the data, it takes time (which is $$$ when it comes to research budgets), and so on. From a theoretical sense, there is a fixed limit on how many variables can contribute to an understanding of what we are trying to predict. Remember that it is best when the predictor or independent variables are independent or unrelated to each other. The problem is that once you get to three or four variables, few things can remain unrelated. Better to be accurate and conservative than to include too many variables and waste money and the power of prediction.

Summary

Prediction is a special case of simple correlations, and it is a very powerful tool for examining complex relationships. This might have been a little more difficult of a chapter than others, but you'll be well served by what you have learned, especially if you can apply it to the research reports and journal articles that you have to read. With the end of lots of chapters about inference, we're to move on to a sample of using statistics when the sample size is very small or when the assumption that the scores are distributed in a normal way is violated.

Time to Practice

1. Chapter 15 Data Set 1 contains the data for a group of participants that took a timed test. The data are the average amount of time the participants took on each item (response time) and the number of guesses it took to get each item correct (number correct).

   a. What is the regression equation for predicting response time from number correct? Use the SLOPE and INTERCEPT functions to create the equation.

   b. What is the predicted response time if the number correct is 8?

   c. What is the difference between the predicted and the actual number correct for each of the predicted response times?

2. Betsy is interested in predicting how many 75-year-olds will develop Alzheimer's disease and is using as predictors level of education and general physical health graded on a scale from 1 to 10. But she is interested in using other predictor variables as well. Answer the following questions.

   a. What criteria should she use in the selection of other predictors? Why?

   b. Name two other predictors that you think might be related to the development of Alzheimer's disease.

   c. With the four predictor variables (level of education and general physical health, and the two new ones that you name), draw the model of what the regression equation would look like.

3. Go to the library and locate three different examples of where linear regression was used in a research study in your area of interest. It's OK if the study contains more than one predictor variable. Answer the following questions for each study.

   a. What is one independent variable? What is the dependent variable?

   b. If there is more than one independent variable, what argument does the researcher make that these variables are independent from one another?

   c. Which of the three studies seems to present the least convincing evidence that the dependent variable is predicted by the independent variable, and why?

4. Here's where you can apply the information in one of this chapter's tips and get a chance to predict a Super Bowl winner! Coach Kent was curious to know if the average number of games won in a year predicts Super Bowl performance (win or lose). The X variable was the average number of games won during the past 10 seasons. The Y variable was whether the team ever won the Super Bowl during the past 10 seasons. Here are the data:

Team	Average Number of Wins Over 10 Years	Ever Win a Super Bowl? (1 = yes and 0 = no)
Savannah Sharks	12	1
Pittsburgh Pelicans	11	0
Williamstown Warriors	15	0
Bennington Bruisers	12	1
Atlanta Angels	13	1
Trenton Terrors	16	0
Virginia Vipers	15	1
Charleston Crooners	9	0
Harrisburg Heathens	8	0
Eaton Energizers	12	1

a. How would you assess the usefulness of the average number of wins as a predictor of whether a team ever won a Super Bowl?

b. What's the advantage of being able to use a categorical variable (such as 1 or 0) as a dependent variable?

c. What other variables might you use to predict the dependent variable, and why would you choose them?

Answers to Practice Questions

1a. The regression equation is $Y' = -.214$(number correct) $+ 17.202$.

1b. $Y' = -.214(8) + 17.202 = 15.49$.

1c.

Time (Y)	# Correct (X)	Y'	Y – Y'
15.5	5	16.13	−1.6
13.4	7	15.70	−2.3
12.7	6	15.92	−3.2
16.4	2	16.77	−0.4
21.0	4	16.35	4.7
13.9	3	16.56	−2.7
17.3	12	15.63	2.7
12.5	5	16.13	−3.6
16.7	4	16.35	0.4
22.7	3	16.56	6.1

2a. The other predictor variables should not be related to any other predictor variable. Only when they are independent of one another can they each contribute unique information to predicting the outcome or dependent variable.

2b. For example, living arrangements (single or in a group) and access to health care (high, medium, or low).

2c. Presence of Alzheimer's disease = (level of education) X_{IV1} + (general physical health) X_{IV2} + (living arrangements) X_{IV3} + (access to health care) X_{IV4} + a.

3. This one you do on your own.

4a. You could compute the correlation between the two variables, which is .204. According to the information in Chapter 5, the magnitude of such a correlation is quite low. You could reach the conclusion that the number of wins is not a very good predictor of whether a team ever won a Super Bowl.

4b. Many variables are categorical by nature (gender, race, social class, and political party) and cannot be measured easily on a scale from 1 to 100, for example. Using categorical variables allows us more flexibility.

4c. Some other variables might be number of All American players, win–loss record of coaches, and home attendance.

16

What to Do When You're Not Normal

Chi-Square and Some Other Nonparametric Tests

Difficulty Scale ☺☺☺☺ (easy)

How much Excel? ⊠ ⊠ ⊠ (lots)

What you'll learn about in this chapter

- What chi-square is and how it can be used
- A bit about the CHIDIS and CHITEST functions
- A brief survey of nonparametric statistics and when and how they should be used

INTRODUCTION TO NONPARAMETRIC STATISTICS

Almost every statistical test that we've covered so far in *Statistics for People Who (Think They) Hate Statistics . . . Excel 2007 Edition* assumes that the data set with which you are working has certain characteristics. For example, one assumption underlying a *t* test between means (be the means independent or dependent) is that the variances of each group are homogeneous, or similar. And the assumptions can be tested. Another assumption of many **parametric statistics** is that the sample is large enough to represent the population. Statisticians have found that it takes a sample size of about 30 to fulfill this assumption. Many of the statistical tests we've covered so far are also robust, or powerful, enough so that even if one of these assumptions is violated, the test is still valid.

But what do you do when the assumptions may be violated? The original research questions are certainly still worth asking and answering. That's when we use **nonparametric statistics** (also called distribution-free statistics). These tests don't follow the same "rules" (meaning they don't require the same assumptions as the parametric tests we've reviewed), but the nonparametrics are just as valuable. The use of nonparametric tests also allows us to analyze data that come as frequencies, such as the number of children in different grades or the percentage of people receiving Social Security.

For example, if we wanted to know whether the number of people who voted for the school voucher in the most recent election is what we would expect by chance, or if there was really a pattern of preference, we would then use a nonparametric technique called **chi-square.**

In this chapter, we cover chi-square, one of the most commonly used nonparametric tests, and provide a brief review of some others just so you can become familiar with some of the nonparametric tests that are available.

INTRODUCTION TO ONE-SAMPLE CHI-SQUARE

Chi-square is an interesting nonparametric test that allows you to determine if what you observe in a distribution of frequencies would be what you would expect to occur by chance. A **one-sample chi-square** or **goodness-of-fit test** includes only one dimension, such as the example you'll see here. A **two-sample chi-square** (also called **test of independence**) includes two dimensions, such as whether preference for the school voucher is independent of political party affiliation and gender. That's left for the next course (or the next edition of this book).

For example, here are data from a sample selected at random from the 1990 census data collected in Sonoma County, California. As you can see, the table organizes information about level of education.

	Level of Education		
No College	Some College	College Degree	Total
25	42	17	84

The question of interest here is whether the number of respondents is equally distributed across all levels of education. To answer this question, the chi-square value (it looks like this: χ^2) was computed and then tested for significance. In this example, the chi-square value is equal to 11.643, which is significant beyond the .05 level. The conclusion is that the number of respondents at the various levels of education for this sample is not equally distributed. In other words, it's not what we would expect by chance, and unequal numbers in level of education appear.

The rationale behind the one-sample chi-square test is that in any set of occurrences, you can easily compute what you would expect by chance. You do this by dividing the total number of occurrences by the number of classes or categories. In our census example above, the observed total number of occurrences was 84. We would expect that, by chance, 84/3 (84, which is the total of all frequencies, divided by 3, which is the total number of categories), or 28, respondents would fall into each of the three categories of level of education.

Then, we look at how different what we expect by chance is from what we observe. If there is no difference between what we expect and what we observe, the chi-square value would be equal to zero.

Let's look more closely at how the chi-square value is computed.

COMPUTING THE CHI-SQUARE TEST STATISTIC

The chi-square test involves a comparison between what is observed and what would be expected by chance. The formula for computing the chi-square value for a one-sample chi-square test is shown in Formula 16.1.

$$\chi^2 = \Sigma \frac{(O - E)^2}{E}, \qquad (16.1)$$

where

χ^2 is the chi-square value,

Σ is the summation sign,

O is the observed frequency, and

E is the expected frequency.

Here are some data we'll use to compute the chi-square value.

Preference for School Voucher			
For	Maybe	Against	Total
23	17	50	90

Here are the famous eight steps to test this statistic.

1. A statement of the null and research hypotheses.

The null hypothesis shown in Formula 16.2 states that there is no difference in the frequency or the proportion of occurrences in each category.

$$H_0: P_1 = P_2 = P_3. \qquad (16.2)$$

The P in the null hypothesis represents the percentage of occurrences in any one category. This null hypothesis states that the percentage of cases in Category 1 (For), Category 2 (Maybe), and Category 3 (Against) is equal. We are using only three categories, but the number could be extended as the situation fits as long as each of the categories is *mutually exclusive*, which means that any one observation cannot be in more than one category. For example, you can't be both male and female. Or, you can't be both for and against the voucher plan at the same time.

The research hypothesis shown in Formula 16.3 states that there is a difference in the frequency or proposition of occurrences in each category.

$$H_0: P_1 \neq P_2 \neq P_3. \qquad (16.3)$$

2. Setting the level of risk (or the level of significance or Type I error) associated with the null hypothesis.

The Type I error rate is set at .05.

3. Selection of the appropriate test statistic.

Any test between frequencies or proportions of mutually exclusive categories (such as For, Maybe, and Against) requires the use of chi-square. The flow chart we have used all along to select the type of statistical test to use is not applicable to nonparametric procedures.

4. Computation of the test statistic value (called the obtained value).

Let's go back to our voucher data from our earlier example and construct a worksheet that will help us compute the chi-square value.

Category	O (observed frequency)	E (expected frequency)	D (difference)	$(O - E)^2$	$(O - E)^2/E$
For	23	30	7	49	1.63
Maybe	17	30	13	169	5.63
Against	50	30	20	400	13.33
Total	90	90			

Here are the steps we took to prepare this worksheet.

1. Enter the categories (Category) of For, Maybe, and Against. Remember that these three categories are mutually exclusive. Any data point can be in only one category at a time.

2. Enter the observed frequency (O), which reflects the data that were collected.

3. Enter the expected frequency (E), which is the total of the observed frequency (90) divided by the number of categories (3), or 90/3 = 30.

4. For each cell, subtract the expected frequency from the observed frequency (D). It does not matter which is subtracted from the other because these values are squared in the next step.

5. Square the observed minus the expected value. You can see these values in the column named $(O - E)^2$.

6. Divide the difference between the observed and the expected frequencies that have been squared, by the expected frequency. You can see these values in the column marked $(O - E)^2/E$.

7. Sum up this last column, and you have the total chi-square value of 20.6.

5. Determination of the value needed for rejection of the null hypothesis using the appropriate table of critical values for the particular statistic.

Here's where we go to Table B.5 (in the appendix) for the list of critical values for the chi-square test.

Our first task is to determine the degrees of freedom (df), which approximates the number of categories in which data have been organized. For this particular test statistic, the degrees of freedom are c − 1, where c equals categories, or 3 − 1 = 2. In other books, or in class, you may see this c − 1 represented as r − 1, for rows minus 1.

Using this number (2) and the level of risk you are willing to take (earlier defined as .05), you can use the chi-square table to look up what the critical value is. It is 5.99. So, at the .05 level, with 2 degrees of freedom, the value needed for rejection of the null hypothesis is 5.99.

6. A comparison of the obtained value and the critical value is made.

The obtained value is 20.6, and the critical value for rejection of the null hypothesis that the frequency of occurrences in Groups 1, 2, and 3 is equal is 5.99.

7. and 8. Decision time!

Now comes our decision. If the obtained value is more extreme than the critical value, the null hypothesis cannot be accepted. If the obtained value does not exceed the critical value, the null hypothesis is the most attractive explanation. In this case, the obtained value exceeds the critical value—it is extreme enough for us to say that the distribution of respondents across the three groups is not equal. Indeed, there is a difference in the frequency of people voting for, maybe, or against when it comes to preference for the school voucher.

TECH TALK A commonly used name for the one-sample chi-square test is goodness of fit. This name suggests the question of how well a set of data "fits" an existing set. The "set" of data is, of course, what you observe. The "fit" part suggests that there is another set of data to which the observed set can be matched. This standard is the set of expected frequencies that is calculated in the course of computing the χ^2 value. If the observed data fit, it's just too close to what you would expect by chance and does not differ significantly. If the observed data do not fit, then what you observed is different from what you would expect.

So How Do I Interpret $\chi^2_{(2)}$ = 20.6, p < .05?

- χ^2 represents the test statistic.
- 2 is the number of degrees of freedom.
- 20.6 is the obtained value using the formula we showed you earlier in the chapter.
- $p < .05$ (the really important part of this little phrase) indicates that the probability is less than 5% on any one test of the null hypothesis that the frequency of votes is equally distributed across all categories by chance alone. Because we defined .05 as our criterion for the research hypothesis being more attractive than the null hypothesis, our conclusion is that there is a significant difference between the two sets of scores.

And Now . . . Using Excel's CHIDIST Function

- Excel doesn't offer any functions that mirror exactly what we did above. There's no function (or ToolPak tool) that computes the value of chi-square for a one-sample test. However, there is a function named **CHIDIST** that will compute the probability of a particular chi-square value if you enter the degrees of freedom and the value of chi-square, such as you see here:

=CHIDIST(20.6, 2).

- In our earlier example, the completed function would look like what you see above and the value that would be returned would be 3.36331E-05, which is .0000336331 (which is pretty darn small). CHIDIST simply provides the probability of an outcome given the degrees of freedom and the value of the statistic χ^2. Keep in mind that you need no raw data to use the value given this function because you already have the chi-square value. Just plug in the value and the degrees of freedom, and you're home free.

But, it's really easy to create your very own formula to compute the chi-square value as you see in Figure 16.1 (look in the formula line at the top of the figure). But, of course, this formula is restricted to three levels of one variable, so if you have more levels, you will have to do a bit of tweaking.

	C6		f_x	=((B2-C2)^2/C2)+((B3-C3)^2/C3)+((B4-C4)^2/C4)	
	A	B	C	D	E	F
1		Observed	Expected			
2	Republican	800	800			
3	Democrat	700	800			
4	Independent	900	800			
5						
6		Chi Square	25			

Figure 16.1 Creating a Formula That Will Compute the Chi-Square Value

TECH TALK There's another chi-square function you should know about, and that is CHITEST. CHITEST asks you for the actual (we called it observed above) and expected values and returns the probability of the resulting chi-square value—not the actual value itself. Just another way to test the general hypothesis of how well a set of data fits what is expected by chance. So, for example, for the data in Figure 16.1, the CHIDIST function requires you to enter the chi-square value of 25 and the number of degrees of freedom and yields that .0000336331. For CHITEST, you enter the range of the actual and the expected values and get the same probability—.0000336331. When to use which? It all depends on the data you have available to be entered into the respective functions.

OTHER NONPARAMETRIC TESTS YOU SHOULD KNOW ABOUT

You may never need a nonparametric test to answer any of the research questions that you propose. On the other hand, you may

Table 16.1   Nonparametric Tests to Analyze Data in Categories and by Ranks

Test Name	When the Test Is Used	A Sample Research Question
To analyze data organized in categories		
McNemar test for significance of changes	To examine "before and after" changes	How effective is a phone call to undecided candidates on their voting for a particular issue?
Fisher's exact test	Computes the exact probability of outcomes in a 2 × 2 table	What is the exact likelihood of getting six heads on a toss of six coins?
Chi-square one-sample test (just like we focused on earlier in this chapter)	To determine if the number of occurrences across categories is random	Did brands Fruities, Whammies, and Zippies each sell an equal number of units during the recent sale?
To analyze data organized by ranks		
Kolmogorov-Smirnov test	To see whether scores from a sample came from a specified population	How representative is a set of judgments of other children of the entire elementary school to which they go?
The sign test, or median test	Used to compare the medians from two samples	Is the median income of people who voted for Candidate A greater than the median income of people who voted for Candidate B?
Mann-Whitney U test	Used to compare two independent samples	Did the transfer of learning, measured by number correct, occur faster for Group A than for Group B?
Wilcoxon rank test	To compare the magnitude as well as the direction of differences between two groups	Is preschool twice as effective as no preschool for helping develop children's language skills?
Kruskal-Wallis one-way analysis of variance	Compares the overall difference between two or more independent samples	How do rankings of supervisors differ between four regional offices?
Friedman two-way analysis of variance	Compares the overall difference between two or more independent samples on more than one dimension	How do rankings of supervisors differ as a function of regional office and gender?
Spearman rank correlation coefficient	Computes the correlation between ranks	What is the correlation between rank in the senior year of high school and rank during the freshman year of college?

very well find yourself dealing with samples that may be very small (or at least fewer than 30) or data that violate some of the important assumptions underlying parametric tests.

TECH TALK

Actually, a primary reason why you may want to use nonparametric statistics is a function of the measurement level of the variable you are assessing. We'll talk more about that in the next chapter, but for now, most data that are categorical and are placed in categories (such as the Sharks and Jets) or that are ordinal and are ranked (1st, 2nd, and 3rd place) call for nonparametric tests of the kind you see in Table 16.1. If that's the case, try nonparametrics on for size. Table 16.1 provides all you need to know about some other nonparametric tests: their name, what they are used for, and a research question that illustrates how each might be used. Keep in mind that the table represents only a few of the many different tests that are available.

Summary

Chi-square is one of many different types of nonparametric statistics that help you answer questions based on data that violate the basic assumptions of the normal distribution or are just too small. These nonparametric tests are a very valuable tool, and even as limited an introduction as this will provide you with some assistance.

Time to Practice

1. Using the following data, test the question that an equal number of Democrats, Republicans, and Independents voted during the most recent election. Test the hypothesis at the .05 level of significance.

Political Affiliation		
Republican	Democrat	Independent
800	700	900

2. Using the following data, test the question that an equal number of boys and girls participate in soccer at the elementary level at the .01 level of significance. What's your conclusion?

Gender	
Boys	Girls
45	55

3. Of the following four research questions, which ones are appropriate for the chi-square test?
   a. The difference between the average scores of two math classes.
   b. The difference between the number of children who passed the math test in Class 1 and the number of children who passed the math test in Class 2.
   c. The number of cars that passed the CRASH Test this year versus last year.
   d. The speed with which a soccer player can run 100 yards compared to the speed of a football player.

Answers to Practice Questions

1. Here's the worksheet for computing the chi-square value:

Category	O (observed frequency)	E (expected frequency)	D (difference)	$(O - E)^2$	$(O - E)^2/E$
Republican	800	800	0	0	0.00
Democrat	700	800	100	10,000	12.50
Independent	900	800	100	10,000	12.50

With 2 degrees of freedom at the .05 level, the critical value needed for rejection of the null hypothesis is 5.99. The obtained value of 25 allows us to reject the null and conclude that there is a significant difference in the numbers of people who voted as a function of political party.

2. Here's the worksheet for computing the chi-square value.

Category	O (observed frequency)	E (expected frequency)	D (difference)	$(O - E)^2$	$(O - E)^2/E$
Boys	45	50	5	25	.5
Girls	55	50	5	25	.5

With 1 degree of freedom at the .01 level of significance, the critical value needed for rejection of the null hypothesis is 6.64. The obtained value of 1.00 means that the null cannot be rejected, and there is no difference between the number of boys and girls who play soccer.

3. Chi-square would be appropriate for Questions b and c because the data that are collected are categorical in nature. Questions a and d deal with data that are continuous (such as average scores).

17

Some Other (Important) Statistical Procedures You Should Know About

Diifficulty Scale ☺☺☺☺ (moderately easy—just some reading and an extension of what you already know)

How much Excel? 📈 (just some mentions)

What you'll learn about in this chapter

- An overview of more advanced statistical procedures and when and how they are used
- Some of the function and data analysis tools you might want to learn more about

Throughout *Statistics for People Who (Think They) Hate Statistics . . . Excel 2007 Edition,* we have covered only a small part of the whole body of statistics. We didn't have room, but more important, at the level at which you are beginning, it's important to keep things simple and direct.

However, that does not mean that in a research article you read or in some discussion in a class, you won't come across other analytical techniques that might be important for you to know about. So, for your edification, here are eight of those techniques, what they do, and examples of studies that used the technique to answer a question. Note that there are always new techniques being developed and new tools that shed light on trends in large data sets, and if you have time, you should definitely try to fit more advanced statistics classes into your coursework. But, for now, the brief summary of what follows should arm you with some knowledge that will at least get you started in the galaxy of advanced statistics.

MULTIVARIATE ANALYSIS OF VARIANCE

You won't be surprised to learn that there are many different renditions of analysis of variance (ANOVA), each one designed to fit a particular "the means of more than two groups being compared" situation. One of these, multivariate analysis of variance (MANOVA), is used when there is more than one dependent variable. So, instead of looking at just one outcome, more than one outcome or dependent variable is used. If the dependent or outcome variables are related to one another (which they usually are—see the Tech Talk note in Chapter 12 about multiple *t* tests), it would be hard to determine clearly the effect of the treatment variable on any one outcome; hence, MANOVA to the rescue. In effect, MANOVA allows you to determine the best combination of dependent variables.

For example, Jonathan Plucker from Indiana University examined gender, race, and grade differences in how gifted adolescents dealt with pressures at school. The MANOVA analysis that he used was a 2 (gender: male and female) × 4 (race: Caucasian, African American, Asian American, and Hispanic) × 5 (grade: 8th through 12th) MANOVA. The *multivariate* part of the analysis was the five subscales of the Adolescent Coping Scale. Using a multivariate technique, the effects of the independent variables (gender, race, and grade) can be estimated for each of the five scales, independent of one another.

Want to know more? Take a look at Plucker, J. A. (1998). Gender, race, and grade differences in gifted adolescents' coping strategies. *Journal for the Education of the Gifted, 21,* 423–436.

REPEATED MEASURES ANALYSIS OF VARIANCE

Here's another kind of analysis of variance. Repeated measures analysis of variance is very similar to any analysis of variance where, if you recall (see Chapter 12), the means of two or more groups are tested for differences. In a repeated measures ANOVA, there is one factor on which participants are tested more than once. That's why it's called repeated, because you repeat the process at more than one point in time on the same factor.

For example, B. Lundy, T. Field, C. McBride, T. Field, and S. Largie examined same-sex and opposite-sex interaction with best friends using juniors who then became seniors in high school. One of their main analyses was ANOVA with three factors: gender (male or female), friendship (same-sex or opposite-sex), and year in high school (junior or senior year). The repeated measure is year in high school, because the measurement was repeated across the same subjects.

Want to know more? Take a look at Lundy, B., Field, T., McBride, C., Field, T., & Largie, S. (1998). Same-sex and opposite-sex best friend interactions among high school juniors and seniors. *Adolescence, 33*(130), 280–289.

What about Excel? The Analysis ToolPak offers the Anova: Two Factor With Replication tool.

ANALYSIS OF COVARIANCE

Here's our last rendition of ANOVA. Analysis of covariance (ANCOVA) is particularly interesting because it basically allows you to equalize initial differences between groups. Let's say you are sponsoring a program to increase running speed and want to compare how fast two groups of athletes can run a 100-yard dash. Because strength is often related to speed, you have to make some correction so that strength does not account for any differences at the end of the program. Rather, you want to see the effects of training with strength removed. You would measure participants' strength before you started the training program and then use ANCOVA to adjust final speed based on initial strength.

Michaela Hynie, John Lyndon, and Ali Tardash from McGill University used ANCOVA in their investigation of the influence of intimacy and commitment on the acceptability of premarital sex and contraceptive use. They used ANCOVA with social acceptability as the dependent variable (in which they were looking for group differences) and ratings of a particular scenario as the covariate. ANCOVA would ensure that differences in social acceptability would be corrected using ratings, so this would be one difference that would be controlled.

Want to know more? See Hynie, M., Lyndon, J., & Tardash, A. (1997). Commitment, intimacy, and women's perceptions of premarital sex and contraceptive readiness. *Psychology of Women's Quarterly, 21*, 447–464.

MULTIPLE REGRESSION

You learned in Chapter 15 how the value of one variable can be used to predict the value of another. Often, social and behavioral sciences researchers look at how more than one variable can predict another. We touched on this in Chapters 5 and 15, and here's more about what is called multiple regression.

For example, it's fairly well established that parents' literacy behaviors (such as having books in the home) are related to how much and how well their children read. So, it would seem quite interesting to look at such variables as parents' age, educational level, literacy activities, and shared reading with children to see what they contribute to early language skills and interest in books.

Paula Lyytinen, Marja-Leena Laakso, and Anna-Maija Poikkeus did exactly that and used stepwise regression analysis to examine the contribution of parental background variables to children's literacy. They found that mothers' literacy activities and mothers' level of education contributed significantly to children's language skills, whereas mothers' age and shared reading did not.

Want to know more? Take a look at Lyytinen, P., Laakso, M. L., & Poikkeus, A. M. (1998). Parental contributions to child's early language and interest in books. *European Journal of Psychology of Education, 3,* 297–308.

What about Excel? The Analysis ToolPak offers the Regression tool.

FACTOR ANALYSIS

Factor analysis is a technique based on how well various items are related to one another and form clusters or factors. Each factor represents several different variables, and factors turn out to be more efficient than individual variables at representing outcomes in certain studies. In using this technique, the goal is to represent those things that are related to one another by a more general name, such as a factor. And the names you assign to these groups of variables called factors is not a willy-nilly process—the names reflect the content and the ideas underlying how they might be related.

For example, David Wolfe and his colleagues at the University of Western Ontario attempted to understand how experiences of maltreatment occurring before children were 12 years old affected peer and dating relationships during adolescence. To do this, the researchers collected data on many different variables and then looked at the relationship between all of them. Those that seemed to contain items that were related (and also belonged to a group that made theoretical sense) were deemed factors, such as the factor named Abuse/Blame in this study. Another factor was named Positive Communication and was made up of 10 different items, all of which were related to each other.

Want to know more? See Wolfe, D. A., Wekerle, C., Reitzel-Jaffe, D., & Lefebvre, L. (1968). Factors associated with abusive relationships among maltreated and non-maltreated youth. *Developmental Psychopathology, 10,* 61–85.

DATA MINING

Data mining is not so much a statistical technique for testing relationships or trends as it is a tool for dealing with the huge data sets that have become commonplace in the social and behavioral sciences. Among other things, it is a way to look for patterns in sets of data. It was first used by the business community when looking at (and for) financial trends, but it is now being used by psychologists, educators, nurses, and others as well. It is very exploratory in nature, sometimes secondary to the main analysis, and it is also used to explore relationships between variables from several different data sets. Data mining is also sometimes referred to as data or knowledge discovery. And, as you might expect, as the technique becomes more popular, new software tools have been developed to facilitate this technique (which frankly was impossible before the advent of computers).

For example, for all you Gen Y students, you well know that online shopping activity has soared over the past 10 years. One study analyzed online buying behavior among college students and used data mining techniques to examine purchases in nine different merchandise categories from 4,688 students. What factors play a role in online shopping (at least for these folks)? The important factors are age, gender, income, car ownership, ability to identify a secure Internet site, and compulsive buying behavior.

Want to know more? See Norum, P. S. (2008). Student Internet purchases. *Family and Consumer Sciences Research Journal, 36*(4), 373–388. Now, where's my iPhone?

PATH ANALYSIS

Here's another statistical technique that examines correlations but allows a bit of a suggestion as to the direction, or causality, in the relationship between factors. Path analysis basically examines the direction of relationships through the postulation of some theoretical relationship between variables and then a test to see if the direction of these relationships is substantiated by the data.

For example, A. Efklides, M. Papadaki, G. Papantonious, and G. Kiosseoglou examined individual feelings of difficulty experienced in the learning of mathematics. To do this, they administered several different types of tests (such as those in the area of cognitive ability) and found that feelings of difficulty are mainly influenced by cognitive (problem-solving) rather than affective (emotional) factors. One of the most interesting uses of path analysis is that a technique called structural equation modeling is used to present the results in a graphical representation of the relationship between all the different factors under consideration. That way, you can actually see what relates to what and with what degree of strength. Then, you can judge how well the data fit the model that was previously suggested. Cool.

Want to know more? Take a look at Efklides, A., Papadaki, M., Papantonious, G., & Kiosseoglou, G. (1998). Individual differences in feelings of difficulty: The case of school mathematics. *European Journal of Psychology of Education, 2,* 207–226.

STRUCTURAL EQUATION MODELING

Structural equation modeling (SEM) is a relatively new technique that has become increasingly popular since it was introduced in the early 1960s. Some researchers feel as if it is an umbrella term for techniques such as regression, factor analysis, and path analysis. Others believe that it stands on its own as an entirely separate approach. It's based on relationships between variables (like the previous three techniques we described).

The major difference between SEM and other advanced techniques such as factor analysis is that SEM is *confirmatory,* rather than *exploratory.* In other words, the researcher is more likely to use SEM to confirm whether a certain model that has been proposed works (meaning the data fit that model). Exploratory techniques set out to discover a particular relationship, with less (but not none) model building beforehand.

For example, Heather Gotham, Kenneth Sher, and Phillip Wood examined the relationships between young adult alcohol use disorders; preadulthood variables (gender, family history of alcoholism, childhood stressors, high school class rank, religious involvement, neuroticism, extraversion, psychoticism); and young adult developmental tasks (baccalaureate degree completion, full-time employment, marriage). Using structural equation modeling techniques, they found that preadulthood variables were more salient predictors of developmental tasks than a diagnosis of having a young adult alcohol use disorder.

Want to know more? Take a look at Gotham, H. J., Sher, K. J., & Wood, P. K. (2003). Alcohol involvement and developmental task completion during young adulthood. *Journal of Studies on Alcohol, 64*(1), 32–42.

Summary

Even though you probably will not be using these more advanced procedures anytime soon, that's all the more reason to know at least something about them, because you will certainly see them mentioned in various research publications and may even hear them mentioned in another class you are taking. And combined with your understanding of the basics (all the chapters in the book up to this one), you can really be confident of having mastered a good deal of important information about basic (and even some intermediate) statistics.

18 A Statistical Software Sampler

Difficulty Scale ☺☺☺☺☺ (a cinch!)

How much Excel? None at all!

What you'll learn about in this chapter

- All about other types of software that allow you to analyze, chart, and better understand your data

You need not be a nerd or anything of the sort to appreciate and enjoy what the various computer programs can do for you in your efforts to learn and use basic statistics. The purpose of this chapter is to give you an overview of some of the more commonly used programs and some of their features, and a quick look at how they work. But before we go into these descriptions, here are some words of advice.

You can find a mega listing of software programs, calculators, and more at Betty C. Jung's home page at http://www.bettycjung .net/Statpgms.htm. There is a gold mine of material here on everything from basic software sites to the basics of statistical software and specialized statistical calculators. This is a great page to explore all the links, and you can never tell what valuable tool you might find.

SELECTING THE PERFECT STATISTICS SOFTWARE

Here are some tried-and-true suggestions for making sure that you get what you want from a stat program.

1. Whether the software program is expensive (like SPSS at about $650) or not so (like EcStatic at about $100), be sure you try it out before you buy it. Almost every stat program listed offers a demo (usually at its Web site) that you can download, and in some cases, you can even ask them to send you a demo version on disk or CD. These versions are often fully featured and last for up to 30 days, which gives you plenty of time to try before you buy.

2. While we're mentioning price, buying it directly from the manufacturer might be the most expensive way to go, especially if you buy outright without inquiring about discounts for students and faculty (what they sometimes call an educational discount). Your school bookstore may offer a discount (or have a license to distribute it free to students), and a mail order company might have even a better deal (again, ask about an educational discount). You can find these sellers' toll-free phone numbers listed in any popular computer magazine. And, remember that prices are always changing.

3. Many of the vendors who produce statistical analysis software offer two flavors. The first is the commercial version, and the second is the academic version. They are usually the same in content but may differ (sometimes dramatically) in price. If you are going for the academic version, be sure that it is the same as the fully featured commercial version, and if not, then ask yourself if you can live with the differences. Why is the academic version so much cheaper? The company hopes that if you are a student, when you graduate, you'll move into some fat-cat job and buy the full version! There may also be a student version where the number of cases or variables you are allowed to enter is limited. Ask, ask, ask.

4. It's hard to know exactly what you'll need before you get started, but some packages come in modules, and you don't

have to buy all of them to get the tools you need for the job you have to do. Read the company's brochures, and call and ask questions.

5. Shareware is another option, and there are plenty of such programs available. Shareware is a method of distributing software so you pay for it only if you like it. Sounds like the honor system, doesn't it? Well, it is. The prices are almost always very reasonable; the shareware is often better than the commercial product; and, if you do pay, you help ensure that the clever author will continue his or her other efforts at delivering new versions that are even better than the one you have.

6. Don't buy any software that does not offer telephone technical support or, at the least, some type of e-mail contact. To test this, call the tech support number (before you buy!) and see how long it takes for them to pick up the phone. If you're on hold for 20 minutes, that may indicate that they don't take tech support seriously enough to get to users' questions quickly. Or, if you e-mail them and never hear back, look for another product.

7. Almost all the big stat packages do the same things—the difference is in the way that they do them. For example, SPSS, Minitab, and JMP all do a nice job of analyzing data and are acceptable. But it's the little things that might make a difference. For example, Minitab allows you to use spaces in the naming of variables while SPSS does not. Go figure.

8. Make sure you have the hardware to run the program you want to use. For example, most software is not limited by the number of cases and variables you want to analyze. The only limit is usually the size of your hard drive, which you'll use to store the data files. And if you have a slow machine (anything less than a Pentium) and less than 256 megabytes of RAM (random access memory), then you're likely to be waiting around and watching that hourglass while your CPU does its thing. Be sure of the hardware you need to run a program before you download the demo. Same goes for the version of your Macintosh operating system or your version of Windows—make sure all is compatible.

WHAT'S OUT THERE

There are more statistical analysis programs available (more than 200) than you would ever need. Here's a listing of some of the most popular and their outstanding features. Remember that many of these do the same thing. If at all possible, as emphasized in the preceding section, try before you buy.

There are a *ton* of these programs available, both free (always and forever), free demos, and those that charge. Have some fun and tool around until you find one that works for you. We'll cover the free programs, and then move on to the software that has some kind of charge associated with it. You can find general descriptions of commercial software at statistics.com (at http://www.statistics.com/content/commsoft/index.php3).

You can also find a mega listing of software programs and links to the home pages of the companies that have created these programs (just select the first letter of the program's name or browse) at http://www.statistics.com/resources/software/commercial/index.php3. You can browse through many different programs and see what appeals.

Your friends at *Statistics for People* galactic headquarters like the following.

The Free Stuff

Free is not bad and, in some cases, may be all that you need. To get a good picture of what's available, visit Free Statistical Software (at http://statpages.org/javasta2.html). Keep in mind that there is the "Completely Free . . ." category (which is no strings attached), as well as the "Free, but . . ." category (which has some strings attached such as "student version," "demos," etc.).

StatCrunch (used to be called WebSTAT)

My favorite? StatCrunch (at http://www.statcrunch.com). What's great about it? First, it's (almost) free—almost $8 per year or $5 for 6 months—and best of all, it's Web based. No need to download software—just "fire it up," enter your data, and calculate what you want. And, because it is Web based, you can use it (and your data), anywhere. Very impressive, convenient, and even fun. In Figure 18.1, you can see a sample screen shot of a data set ready to undergo analysis.

SSP (Smith's Statistical Package)

SSP is a simple stat package for Mac and Windows that does lots of SPSS (the big dog on the block) things such as calculate basic summaries; prepare charts; compare means; and do ANOVAs, chi-square tests, and regression. You can see a simple *t* test in Figure 18.2.

Figure 18.1   StatCrunch: Statistics on the Web

Need more information? Go to http://www.economics.pomona
.edu/StatSite/SSP.html.

Difference-In-Means Test	✕

A statistical test of the null hypothesis that two population means are equal is based on the t value, which measures how many standard errors the difference in the sample means is from 0. This standard error can be estimated without assuming that the population standard deviations are equal, or by assuming that they are equal and using a pooled variance.

Group, row 1 to row 20, number of actual observations:	20
Number of missing observations:	0
The mean of the first sample:	1.5000
The standard deviation of the first sample:	0.5130
Score, row 1 to row 20, number of actual observations:	20
Number of missing observations:	0
The mean of the second sample:	6.1000
The standard deviation of the second sample:	1.8610
The t value with 21.8710 degrees of freedom:	10.6570
The two-sided p value:	0.0000000004
Assuming equal population standard deviations, the pooled variance:	1.8632
The t value with 38 degrees of freedom:	10.6570
The two-sided p value:	5.65598×10^{-13}

Figure 18.2   A Simple *t* Test Done in SSP

Time to Pay

OK, enough free stuff. Don't forget—first see what the free stuff can do, and if it is not enough, then look at these.

JMP

JMP (now in version 7) is billed as the "statistical discovery software." It operates on Mac, Windows, and also Linux platforms and is software that "links statistics with graphics to interactively explore, understand, and visualize data." One of JMP's features is to present a graph accompanying every statistic so you can always see the results of the analysis both as statistical text and as a graphic. And all this is done automatically without you requesting it.

Need more information? Go to http://www.jmp.com/software or give them a call at 1–877–594–6567.

Cost: $595 for a single academic license.

Minitab

This is one of the first programs available for the personal computer, and it is now in Version 16 (it's been around a while!), which means that it's seen its share of changes over the years in response to users' needs. Some of the more outstanding features of this new version are

- Customizable menus and toolbars;
- StatGuide™, which helps explain output;
- ReportPad™, which is a report generator;
- Online tutorials;
- User-editable profiles;
- New graphics engine; and
- Lots of new multivariate tools.

Need more information? Go to http://www.minitab.com or call 1–800–448–3555.

Cost: $1,195 for the full mega version, $100 for the academic version (cheaper versions are available that are time limited as well).

STATISTICA

StatSoft (at http://www.statsoft.com) offers a collection of products for Windows, including STATISTICA 8, and for anyone with a Web browser, WebSTATISTICA. Some of the features that are particularly nice about these powerful programs are the self-prompting dialog boxes and templates (you click OK and STATISTICA tells you what to enter); the customization of the interface; easy integration with other programs; STATISTICA Visual Basic, which allows you to access more than 11,000 functions and use this development environment to design special applications; and the ability to use macros to automate tasks.

A nice bonus at the Web site is an Electronic Statistics Textbook (or EST), which you can download in its entirety (have patience, because it can take up to 30 minutes depending on the speed of your Internet connection), or you can buy a hard copy for a mere $80.

Need more information? Go to http://www.statsoftinc.com or call 1–918–749–1119.

Cost: $250 for the academic version.

SPSS

SPSS may be the most popular big-time statistical package in use. It comes with a variety of different modules that cover all aspects of statistical analysis, including both basic and advanced statistics, and a version exists for several platforms (Windows XP/Vista, Mac OS X [Tiger and Leopard], and Linux).

One of the nicer new features included in Version 17 is new chart types and templates that are easier to work with and use. Figure 18.3 shows a simple bar graph created in SPSS from a simple data set telling you how many males and females are in a group. SPSS also features a powerful report writer.

Need more information? Go to http://www.spss.com or give them a call at 1–312–651–3000.

Cost: $639 for the academic Windows version, and around the same for the Mac version (which looks just like the Windows version). They offer a student version as well, so check on the Web site or call the SPSS folks to find out if it is available.

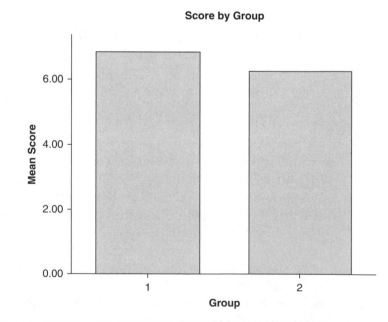

Figure 18.3 An SPSS Graph

SYSTAT

SYSTAT tends to be used by researchers in biological and physical sciences, whereas the social and behavioral sciences researchers like SPSS (although the SYSTAT people are making an effort to appeal to the social and behavioral sciences people in the newest release, version 12). It's Windows only (with a Mac version in the works) and supports a strong command language so that analysis can be fine-tuned to users' needs. The latest version also sports a new interface and highly customizable menus. The beginner can use this stuff, but it's more appropriate for the more advanced student or professional.

Need more information? Go to http://systat.com or call them at 1–877–797–8280.

Cost: $1,299 for the commercial version, but you can download MYSTAT, which is a free version of SYSTAT and does lots of what you will need.

STATISTIX for Windows

Version 9 of STATISTIX (only for Windows) is as powerful as the other programs described here but also offers a menu-driven interface that makes it particularly easy to learn to use, offers free technical support, and—ready for this?—a real, 400-page paper manual. And when you call technical support, you talk with the actual programmers, who know what they're talking about (my question was answered in 10 seconds!). All around, a good deal.

Need more information? Go to http://www.statistix.com or call 1–800–933–7879.

Cost: $395 for the academic version.

EcStatic

The goal of the people at Someware in Vermont software is to "provide intelligently crafted, easy-to-use statistical and graphing software at reasonable prices." They do this and more. EcStatic is a steal for the money. It is the least expensive of any of the programs that can really perform, and you certainly get much more than you pay for compared to the huge programs described above. And if you think that this program is missing anything, take a look at the following list of some of the things it can do:

- Analysis of variance
- Breakdown
- Conversion of scores
- Correlation
- Cross-tabulation and chi-square

- Frequency distributions and histograms
- Nonparametric statistics
- Regression
- Scatterplot
- Summary statistics
- Transformations
- *t* tests

Download a trial copy now and have some fun.

Need more information? Go to http://www.somewareinvt.com.

Cost: $99.95 to download, with a discount on 10 or more ($69.95)—tell your instructor!

Summary

That's the end of Part IV and just about the end of *Statistics for People Who (Think They) Hate Statistics . . . Excel 2007 Edition*. But read on! The next chapter includes the ten best Internet sites in the universe for information about statistics, followed by Chapter 20, the ten commandments of data collection. Have fun with both of these.

PART V

Ten Things You'll Want to Know and Remember

Snapshots

"The Internet's got him ... Reboot! REBOOT!"

19

The Ten (or More) Best (and Most Fun) Internet Sites for Statistics Stuff

I f you're not yet using the Internet as a part of your learning and research activities, you are missing out on an extraordinary resource. Today, more than ever, students, researchers, and others certainly are taking advantage of this vast resource, but there's still some reluctance on the part of newbies.

What's available on the Internet will not make up for a lack of studying or motivation—nothing will do that—but you can certainly find a great deal of information that will enhance your whole college experience. And this doesn't even begin to include all the fun you can have!

So, now that you're a certified novice statistician, here are some Internet sites that you might find very useful should you want to learn more about statistics.

Although the locations of Web sites on the Internet are more stable than ever, they still can change frequently. The URL (that http:// thing that precedes all Internet addresses called the uniform resource locator) that worked today might not work tomorrow. If you don't get to where you want to go, use Google or some other search engine and search the name—perhaps there's a new URL or other Web address that works.

HOW ABOUT STUDYING STATISTICS IN STOCKHOLM?

The World Wide Web Virtual Library: Statistics is the name of the page, but the one-word title is misleading because the site (from the good people at the University of Florida at http://www.stat.ufl.edu/vlib/statistics.html) includes information on just about every facet of the topic, including data sources, job announcements, departments, divisions and schools of statistics (a huge description of programs all over the world), statistical research groups, institutes and associations, statistical services, statistical archives and resources, statistical software vendors and software, statistical journals, mailing list archives, and related fields. Tons of great information is available here. Make it a stop along the way.

CALCULATORS GALORE!

Want to draw a histogram? How about a table of random numbers? A sample size calculator? The Statistical Calculators page at http://www.stat.ucla.edu/calculators and from the very good people at UCLA has just about every type (more than 15) of calculator and table you could need. Enough to carry you through any statistics course that you might take and even more. Very, very cool stuff.

For example, you can click on the Random Permutations link and complete the two boxes (as you see in Figure 19.1 for two random permutations of 100 integers), and you get the number of permutations you want. This is very handy when you need a table of random numbers for a specific number of participants so you can assign them to groups. You can find another assortment of calculators at http://www.bettycjung.net/Statpgms.htm as well.

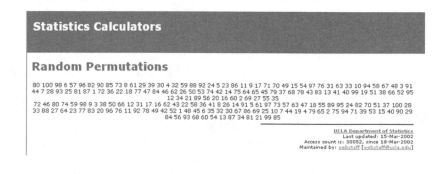

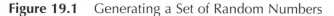

Figure 19.1 Generating a Set of Random Numbers

WHO'S WHO AND WHAT'S HAPPENED

The History of Statistics page located at http://www.Anselm.edu/homepage/jpitocch/biostatshist.html contains portraits and bibliographies of famous statisticians and a time line of important contributions to the field of statistics. So, names like Bernoulli, Galton, Fisher, and Spearman pique your curiosity? How about the development of the first test between two averages during the early 20th century? It might seem a bit boring until you have a chance to read about the people who make up this field and their ideas—in sum, pretty cool ideas and pretty cool people.

IT'S ALL HERE

SurfStat Australia (http://surfstat.anu.edu.au/surfstat-home/surfstat-main.html) is the online component of a basic stat course taught at the University of Newcastle, Australia, but has grown far beyond just the notes originally written by Annette Dobson in 1987 and updated over several years' use by Anne Young, Bob Gibberd, and others. Among other things, SurfStat contains a complete interactive statistics text. Besides the text, there are exercises, a list of other statistics sites on the Internet, and a collection of Java applets (cool little programs you can use to work with different statistical procedures).

HYPERSTAT

This online tutorial with 18 lessons, at http://www.davidmlane.com/hyperstat/index.html, offers nicely designed and user-friendly coverage of the important basic topics. What we really liked about the site was the glossary, which uses hypertext to connect different concepts to one another. For example, in Figure 19.2, you can see the definition of descriptive statistics also linked to other glossary terms, such as mean, standard deviation, and box plot. Click on any of those and zap! You're there.

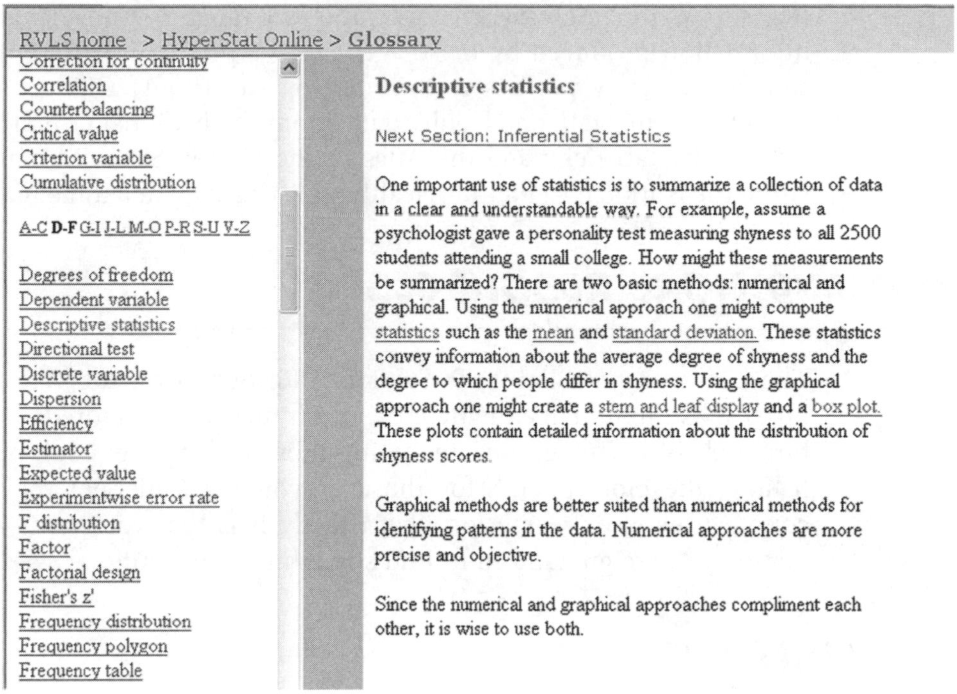

Figure 19.2 Sample HyperStat Screen

DATA? YOU WANT DATA?

There are data all over the place, ripe for the picking. Here are just a few. What to do with these? Download them to be used as examples in your work or as examples of analysis that you might want to do, and you can use these as a model.

- Statistical Reference Datasets at http://www.itl.nist.gov/div898/strd
- United States Census Bureau (a huge collection and a goldmine of data) at http://factfinder.census.gov/servlet/DatasetMainPage Servlet?_program=DEC&_submenuId=datasets_0&_lang=en
- The Data and Story Library (http://lib.stat.cmu.edu/DASL/) with great annotations about the data (look for the stories link)
- Tons of economic data sets at Growth Data Sets (at http://www.bris.ac.uk/Depts/Economics/Growth/datasets.htm)

Then, even more data sets are available through the federal government (besides the Census Bureau and others highlighted above). Your tax money supports it, so why not use it? For example, there's FEDSTATS (at http://www.fedstats.gov), where more than 70 agencies in the U.S. government produce statistics of interest to the public. The Federal Interagency Council on Statistical Policy maintains this site to provide easy access to the full range of statistics and information produced by these agencies for public use. Here you can find country profiles contributed by the (boo!) CIA; public school student, staff, and faculty data (from the National Center for Education Statistics); and the Atlas of the United States Mortality (from the National Center for Health Statistics). What a ton of data!

MORE AND MORE RESOURCES

The University of Michigan's Statistical Resources on the Web (at http://www.lib.umich.edu/govdocs/stats.html) has hundreds and hundreds of resource links, including those to banking; book publishing; the elderly; and, for those of you with allergies, pollen count. Browse, or search for what exactly it is that you need—no matter, you are guaranteed to find something interesting.

PLAIN, BUT FUN

At http://mathforum.org/workshops/sum96/data.collections/datalibrary/data.set6.html, you can find a data set including the 1994 National League Baseball Salaries or the data on TV, Physicians, and Life Expectancy. Nothing earth-shaking, just fun to download and play with.

ONLINE STATISTICAL TEACHING MATERIALS

You're so good at this statistics stuff that you might as well start helping your neighbor and colleague in class. If that's the case, turn to Web Links Concerning Statistics Teaching at http://www.math .unb.ca/~knight/webstatx.htm for tons of content and links, glossaries and fun (yes, fun).

AND OF COURSE YOUTUBE . . .

Yes, you can now find stat stuff on YouTube in the form of Statz Rappers (at http://www.youtube.com/watch?v=JS9GmU5hr5w), a group of talented young men and women who seem to be having a great time making just a bit of fun of their stat course—a very fitting last stop on this path of what the Internet holds for those interested in exploring statistics.

20 The Ten Commandments of Data Collection

Now that you know how to analyze data, you would be well served to hear something about collecting them. The data collection process can be a long and rigorous one, even if it involves only a simple, one-page questionnaire given to a group of students, parents, patients, or voters. The data collection process may very well be the most time-consuming part of your project. But as many researchers do, this time is also used to think about the upcoming analysis and what it will entail.

Here they are: the ten commandments for making sure your data get collected in a way that they are usable. Unlike the original Ten Commandments, these should not be carved in stone (because they can certainly change), but if you follow them, you can avoid lots of aggravation.

Commandment 1. As you begin thinking about a research question, also begin thinking about the type of data you will have to collect to answer that question. Interview? Questionnaire? Paper and pencil? Find out how other people have done it in the past by reading the relevant journals in your area of interest and consider doing what they did.

Commandment 2. As you think about the type of data you will be collecting, think about where you will be getting the data. If you are using the library for historical data or accessing files of data that have already been collected, such as census data (available through the U.S. Census Bureau and some online sites), you will have few logistical problems. But what if you want to assess the interaction between newborns and their parents? The attitude of teachers toward unionizing? The age at which people over 50 think they are old? All of these questions involve needing people to provide the answers, and finding people can be tough. Start now.

Commandment 3. Make sure that the data collection forms you use are clear and easy to use. Practice on a set of pilot data so you can make sure it is easy to go from the original scoring sheets to the data collection form.

Commandment 4. Always make a duplicate copy of the data file, and keep it in a separate location. Keep in mind that there are two types of people: those who have lost their data and those who will. Keep a copy of data collection sheets in a separate location. If you are recording your data as a computer file, such as a spreadsheet, be sure to make a backup!

Commandment 5. Do not rely on other people to collect or transfer your data unless you have personally trained them and are confident that they understand the data collection process as well as you do. It is great to have people help you, and it helps keep the morale up during those long data collection sessions. But unless your helpers are competent beyond question, you could easily sabotage all your hard work and planning.

Commandment 6. Plan a detailed schedule of when and where you will be collecting your data. If you need to visit three schools and each of 50 children needs to be tested for a total of 10 minutes at each school, that is 25 hours of testing. That does not mean you can allot 25 hours from your schedule for this activity. What about travel from one school to another? What about the child who is in the bathroom when it is his turn, and you have to wait 10 minutes until he comes back to the classroom? What about the day you show up and Cowboy Bob is the featured guest . . . and on and on. Be prepared for anything, and allocate 25% to 50% more time in your schedule for unforeseen happenings.

Commandment 7. As soon as possible, cultivate possible sources for your subject pool. Because you already have some knowledge in your own discipline, you probably also know of people who work with the type of population you want or who might be able to help you gain access to these samples. If you are in a university community, it is likely that there are hundreds of other people competing for the same subject sample that you need. Instead of competing, why not try a more out-of-the-way (maybe 30 minutes away) school district or social group or civic organization or hospital, where you might be able to obtain a sample with less competition?

Commandment 8. Try to follow up on subjects who missed their testing session or interview. Call them back and try to reschedule.

Once you get in the habit of skipping possible participants, it becomes too easy to cut the sample down to too small a size. And you can never tell—the people who drop out might be dropping out for reasons related to what you are studying. This can mean that your final sample of people is qualitatively different from the sample with which you started.

Commandment 9. Never discard the original data, such as the test booklets, interview notes, and so forth. Other researchers might want to use the same database, or you may have to return to the original materials for further information.

And number 10? Follow the previous 9. No kidding!

APPENDIX A

Excel-erate Your Learning

All You Need to Know About Excel

This is not a book that teaches you how to use Excel, although it is filled with lots of terrific bits and pieces about how to do exactly that.

So, if you have little experience with Excel, this is the place to start. We've listed 50 of the most important and useful beginning tasks you should master to have the level of competence you need to breeze through using Excel as a tool for understanding simple data analysis.

Our advice is to open a new or practice worksheet where no damage could be done and enter a good deal of data. Then, practice, practice, practice. And, here's the best tip you'll ever get about Excel. Try everything—have no fear—no damage can be done and you'll only learn and have fun doing it.

A big change from the last version in Excel is the Office Button that looks like this ⬤. Lots and lots of operations, simple and complex, begin here in the upper left-hand corner of the Excel screen, and so much of what you need is quite handy.

One other big change that is a feature that most users welcome and like, and that is the creation, and use, of the Quick Access Toolbar.

You could always customize Toolbars in Excel, but now you can create kind of a summary toolbar right above the Excel Ribbon (which is the set of tabs at the top of the screen). To add an icon to the Quick Access Toolbar, follow these steps.

1. Click the small drop-down arrow to the right of the Office Button.

2. Select the More Commands option.

3. Select the All Commands option from the Choose commands from the drop-down menu.

4. Highlight the command you want to use on the left, and click the Add button to add it to the list on the right.

5. Click OK and the button for the command will appear in the Quick Access Toolbar. No more hunting through tabs and menus—just click what you want. Figure A.1 shows you a sample Quick Access Toolbar with commands for New, Open, Save, Print, AutoFit Column Width, Font Color, and Sort. Pretty cool.

Figure A.1   A Simple Quick Access Toolbar: Customizable and a Great Time Saver

JUST THE BASICS

1. To create a new worksheet, click the Office Button and then click New, click the New button, or use the Ctrl+N key combination.

2. To open an existing worksheet, click the Office Button and then click Open, click the Open button, or use the Ctrl+O key combination.

3. To save a worksheet, click the Office Button and then click Save, click the Save button, or use the Ctrl+S key combination.

4. To save a worksheet under a different name, click the Office Button and then click Save As.

5. Need help? Press the F1 key or click the question mark on the right side of the spreadsheet.

ENTERING DATA IN A WORKSHEET

1. To enter data in a worksheet, click on the cell, type in the data, and press Enter.

2. To enter data in a series, enter the first value in the series, enter the second value, highlight both of them, and drag the latest value in the series to the last cell.

3. Screwed up? Use the Ctrl+Z key combination to reverse the last operation you asked Excel to perform, be it simple data entry or a huge calculation.

FINDING AND REPLACING DATA

1. To find data in a worksheet, click the Home tab and click Find or use the Ctrl+F key combination, enter the data you want to find, and click Find Next.

2. To find and replace data, click the Home tab and click Replace or use the Ctrl+H key combination, enter the data you wish to find, and click Replace.

EDITING A WORKSHEET

1. To select a cell, click on that cell. To select the entire worksheet, click on the empty box next to column A and row 1.

2. To edit a cell, click on that cell, press the F2 key, and make the edits in the cell or the formula bar.

3. To select a row or column, click on the heading for that row (a number) or that column (a letter).

4. To insert a row or column, right click on the column letter or row number and click. You can also use the Insert and Delete buttons on the Home tab.

5. To copy data from one cell or a range of cells, highlight the data to be copied, right click → Copy, click on the cell in which you want the data copied, and right click → Paste.

6. To copy data from one cell or a range of cells, you can also highlight the data to be copied, place the mouse pointer on the cell border until it appears as a hand, press the Ctrl key, and drag the data to the new location.

7. To replace data, click on the cell, type the new data, and press Enter.

8. To cut data from one cell or a range of cells, highlight the data to be cut, right click → Cut or press the Delete key.

9. To format a number, select the cell, right click Format Cells, and click the Number tab. Select the format and click OK.

WORKING WITH A WORKSHEET APPEARANCE

1. To make rows or columns adjust to fit the data in a worksheet, highlight the entire worksheet, then click the Home tab, then click the Format option, and then click AutoFit Column Width or AutoFit Row Height.

2. To change the format of data, highlight the cells you want to change, right click Format Cells, and make the changes.

3. To change the default font, click the Microsoft button, click the Personalize category, and select the font and the font size.

4. To add borders to a cell, highlight the cells you want to change, then click the border change you want on the drop down menu of the Border button.

5. To add shading to a cell, click the Home tab, then click the shading tool, and select from the drop down menu.

6. To change the margins of a worksheet, click the Microsoft Office button, place the mouse pointer on Print, click Print Preview, click the Page Setup button, select the Margins tab, and adjust the margins.

7. To add a header or a footer to a worksheet, click the Insert tab, then the Header and Footer tab, enter the text for the header and/or footer, and click on any area outside of the header or footer.

8. To show or hide gridlines, click the View tab and then click the Gridlines box.

9. To bold data, highlight it and click the Bold button on the Home tab.

10. To italicize data, highlight it and click the Italics button on the Home tab.

11. To change the alignment of text, highlight the left, center, or right alignment button on the Home tab.

12. To change the number of decimal places in a number, click on the Increase Decimal or Decrease Decimal icons on the Home tab.

13. To add row borders to a cell or set of cells, click the Borders button on the Home tab and make the adjustments you want.

WORKING WITH CELLS AND VALUES

1. To name a range of cells, select the cells you want to name, right click, click Name a Range, provide a range, and click OK.

2. To create a formula, click the cell in which you want the results of the formula returned, type an equal sign (=), type the formula, and press Enter.

3. To enter a function, click the cell in which you want the results of the function returned, click the Formula tab, click the group of functions containing the function you want to use, and then select the function.

4. To enter the sum function, click on the summation sign on the Home tab and click Sum.

5. To sort data, select the cells you want to sort, click the Data tab, and select sort by Ascending (low to high) or Descending (high to low).

6. To hide columns or rows, select the column or row you want to hide, click the View tab, and click Hide.

7. To unhide columns or rows, select the column or row adjacent to the one that is hidden, click the View tab, and click Unhide.

USING THE ANALYSIS TOOLPAK

1. To use the Analysis ToolPak, click the Data tab, then click Data Analysis, and double click the analysis you want to do.

CREATING AND USING CHARTS

1. To create a chart, highlight the data including the column and row labels, and click the Insert tab and the type of chart you want to create.

2. To resize chart, drag on the chart handles horizontally, vertically, or diagonally.

3. To change chart labels and information, click the Layout tab and then the Chart Title tab.

4. To print only a chart, click on the chart to highlight it, click the Office button, and then click Print.

5. To change the format of a chart element, click on that element, make the changes you want, and then click OK.

6. To copy a chart to another application, highlight it, press the Ctrl+A key combination, press Ctrl+C, and go to the new application, and press Ctrl+V.

PRINTING WORKSHEETS

1. To print a worksheet, click the Office Button and then click Print.

2. To preview a worksheet, click the Office Button, then click Print, and then select Print Preview.

3. To zoom in on part of a worksheet, click the Print Preview icon and then click Zoom.

4. To print nonadjacent sets of cells in a worksheet, hold down the Ctrl button as they are selected and then print them.

APPENDIX B

Tables

TABLE B.1: AREAS BENEATH THE NORMAL CURVE

How to use this table:

1. Compute the z score based on the raw score and the mean of the sample.

2. Read to the right of the z score to determine the percentage of area underneath the normal curve or the area between the mean and computed z score.

TABLE B.1 Areas Beneath the Normal Curve

z score	Area Between the Mean and the z score	z score	Area Between the Mean and the z score	z score	Area Between the Mean and the z score	z score	Area Between the Mean and the z score	z score	Area Between the Mean and the z score	z score	Area Between the Mean and the z score	z score	Area Between the Mean and the z score	z score	Area Between the Mean and the z score
0.00	0.00	0.50	19.15	1.00	34.13	1.50	43.32	2.00	47.72	2.50	49.38	3.00	49.87	3.50	49.98
0.01	0.40	0.51	19.50	1.01	34.38	1.51	43.45	2.01	47.78	2.51	49.40	3.01	49.87	3.51	49.98
0.02	0.50	0.52	19.85	1.02	34.61	1.52	43.57	2.02	47.83	2.52	49.41	3.02	49.87	3.52	49.98
0.03	1.20	0.53	20.19	1.03	34.85	1.53	43.70	2.03	47.88	2.53	49.43	3.03	49.88	3.53	49.98
0.04	1.60	0.54	20.54	1.04	35.08	1.54	43.82	2.04	47.93	2.54	49.45	3.04	49.88	3.54	49.98
0.05	1.99	0.55	20.88	1.05	35.31	1.55	43.94	2.05	47.98	2.55	49.46	3.05	49.89	3.55	49.98
0.06	2.39	0.56	21.23	1.06	35.54	1.56	44.06	2.06	48.03	2.56	49.48	3.06	49.89	3.56	49.98
0.07	2.79	0.57	21.57	1.07	35.77	1.57	44.18	2.07	48.08	2.57	49.49	3.07	49.89	3.57	49.98
0.08	3.19	0.58	21.90	1.08	35.99	1.58	44.29	2.08	48.12	2.58	49.51	3.08	49.90	3.58	49.98
0.09	3.59	0.59	22.24	1.09	36.21	1.59	44.41	2.09	48.17	2.59	49.52	3.09	49.90	3.59	49.98
0.10	3.98	0.60	22.57	1.10	36.43	1.60	44.52	2.10	48.21	2.60	49.53	3.10	49.90	3.60	49.98
0.11	4.38	0.61	22.91	1.11	36.65	1.61	44.63	2.11	48.26	2.61	49.55	3.11	49.91	3.61	49.98
0.12	4.78	0.62	23.24	1.12	36.86	1.62	44.74	2.12	48.30	2.62	49.56	3.12	49.91	3.62	49.98
0.13	5.17	0.63	23.57	1.13	37.08	1.63	44.84	2.13	48.34	2.63	49.57	3.13	49.91	3.63	49.98
0.14	5.57	0.64	23.89	1.14	37.29	1.64	44.95	2.14	48.38	2.64	49.59	3.14	49.92	3.64	49.98
0.15	5.96	0.65	24.54	1.15	37.49	1.65	45.05	2.15	48.42	2.65	49.60	3.15	49.92	3.65	49.98
0.16	6.36	0.66	24.86	1.16	37.70	1.66	45.15	2.16	48.46	2.66	49.61	3.16	49.92	3.66	49.98
0.17	6.75	0.67	25.17	1.17	37.90	1.67	45.25	2.17	48.50	2.67	49.62	3.17	49.92	3.67	49.98
0.18	7.14	0.68	25.49	1.18	38.10	1.68	45.35	2.18	48.54	2.68	49.63	3.18	49.93	3.68	49.98
0.19	7.53	0.69	25.80	1.19	38.30	1.69	45.45	2.19	48.57	2.69	49.64	3.19	49.93	3.69	49.98
0.20	7.93	0.70	26.11	1.20	38.49	1.70	45.54	2.20	48.61	2.70	49.65	3.20	49.93	3.70	49.99
0.21	8.32	0.71	26.42	1.21	38.69	1.71	45.64	2.21	48.64	2.71	49.66	3.21	49.93	3.71	49.99
0.22	8.71	0.72	26.73	1.22	38.88	1.72	45.73	2.22	48.68	2.72	49.67	3.22	49.94	3.72	49.99
0.23	9.10	0.73	27.04	1.23	39.07	1.73	45.82	2.23	48.71	2.73	49.68	3.23	49.94	3.73	49.99

(Continued)

TABLE B.1 (Continued)

z score	Area Between the Mean and the z score	z score	Area Between the Mean and the z score	z score	Area Between the Mean and the z score	z score	Area Between the Mean and the z score	z score	Area Between the Mean and the z score	z score	Area Between the Mean and the z score	z score	Area Between the Mean and the z score	z score	Area Between the Mean and the z score
0.24	9.48	0.75	27.34	1.24	39.25	1.74	45.91	2.24	48.75	2.74	49.69	3.24	49.94	3.74	49.99
0.25	9.99	0.76	27.64	1.25	39.44	1.75	45.99	2.25	45.78	2.75	49.70	3.25	49.94	3.75	49.99
0.26	10.26	0.77	27.94	1.26	39.62	1.76	46.08	2.26	48.81	2.76	49.71	3.26	49.94	3.76	49.99
0.27	10.64	0.78	28.23	1.27	39.80	1.77	46.16	2.27	48.84	2.77	49.72	3.27	49.94	3.77	49.99
0.28	11.03	0.79	28.52	1.28	39.97	1.78	46.25	2.28	48.87	2.78	49.73	3.28	49.94	3.78	49.99
0.29	11.41	0.80	28.81	1.29	40.15	1.79	46.33	2.29	48.90	2.79	49.74	3.29	49.94	3.79	49.99
0.30	11.79	0.81	29.10	1.30	40.32	1.80	46.41	2.30	48.93	2.80	49.74	3.30	49.95	3.80	49.99
0.31	12.17	0.82	29.39	1.31	40.49	1.81	46.49	2.31	48.96	2.81	49.75	3.31	49.95	3.81	49.99
0.32	12.55	0.83	29.67	1.32	40.66	1.82	46.56	2.32	48.98	2.82	49.76	3.32	49.95	3.82	49.99
0.33	12.93	0.84	29.95	1.33	40.82	1.83	46.64	2.33	49.01	2.83	49.77	3.33	49.95	3.83	49.99
0.34	13.31	0.85	30.23	1.34	40.99	1.84	46.71	2.34	49.04	2.84	49.77	3.34	49.95	3.84	49.99
0.35	13.68	0.86	30.51	1.35	41.15	1.85	46.78	2.35	49.06	2.85	49.78	3.35	49.96	3.85	49.99
0.36	14.06	0.87	30.78	1.36	41.31	1.86	46.86	2.36	49.09	2.86	49.79	3.36	49.96	3.86	49.99
0.37	14.43	0.88	31.06	1.37	41.47	1.87	46.93	2.37	49.11	2.87	49.79	3.37	49.96	3.87	49.99
0.38	14.80	0.89	31.33	1.38	41.62	1.88	46.99	2.38	49.13	2.88	49.80	3.38	49.96	3.88	49.99
0.39	15.17	0.90	31.59	1.39	41.77	1.89	47.06	2.39	49.16	2.89	49.81	3.39	49.96	3.89	49.99
0.40	15.54	0.91	31.86	1.40	41.92	1.90	47.13	2.40	49.18	2.90	49.81	3.40	49.97	3.90	49.99
0.41	15.91	0.92	32.12	1.41	42.07	1.91	47.19	2.41	49.20	2.91	49.82	3.41	49.97	3.91	49.99
0.42	16.28	0.93	32.38	1.42	42.22	1.92	47.26	2.42	49.22	2.92	49.82	3.42	49.97	3.92	49.99
0.43	16.64	0.94	32.64	1.43	42.36	1.93	47.32	2.43	49.25	2.93	49.83	3.43	49.97	3.93	49.99
0.44	17.00	0.95	32.89	1.44	42.51	1.94	47.38	2.44	49.27	2.94	49.84	3.44	49.97	3.94	49.99
0.45	17.36	0.96	33.15	1.45	42.65	1.95	47.44	2.45	49.29	2.95	49.84	3.45	49.98	3.95	49.99
0.46	17.72	0.97	33.40	1.46	42.79	1.96	47.50	2.46	49.31	2.96	49.85	3.46	49.98	3.96	49.99
0.47	18.08	0.98	33.65	1.47	42.92	1.97	47.56	2.47	49.32	2.97	49.85	3.47	49.98	3.97	49.99
0.48	18.44	0.99	33.89	1.48	43.06	1.98	47.61	2.48	49.34	2.98	49.86	3.48	49.98	3.98	49.99
0.49	18.79	1.00	34.13	1.49	43.19	1.99	47.67	2.49	49.36	2.99	49.86	3.49	49.98	3.99	49.99

TABLE B.2: T VALUES NEEDED FOR REJECTION OF THE NULL HYPOTHESIS

How to use this table:

1. Compute the t value test statistic.

2. Compare the obtained t value to the critical value listed in this table. Be sure you have calculated the number of degrees of freedom correctly and you have selected an appropriate level of significance.

3. If the obtained value is greater than the critical or tabled value, the null hypothesis (that the means are equal) is not the most attractive explanation for any observed differences.

4. If the obtained value is less than the critical or table value, the null hypothesis is the most attractive explanation for any observed differences.

TABLE B.2 *t* Values Needed for Rejection of the Null Hypothesis

df	One-Tailed Test			df	Two-Tailed Test		
	0.10	*0.05*	*0.01*		*0.10*	*0.05*	*0.01*
1	3.078	6.314	31.821	1	6.314	12.706	63.657
2	1.886	2.92	6.965	2	2.92	4.303	9.925
3	1.638	2.353	4.541	3	2.353	3.182	5.841
4	1.533	2.132	3.747	4	2.132	2.776	4.604
5	1.476	2.015	3.365	5	2.015	2.571	4.032
6	1.44	1.943	3.143	6	1.943	2.447	3.708
7	1.415	1.895	2.998	7	1.895	2.365	3.5
8	1.397	1.86	2.897	8	1.86	2.306	3.356
9	1.383	1.833	2.822	9	1.833	2.262	3.25
10	1.372	1.813	2.764	10	1.813	2.228	3.17
11	1.364	1.796	2.718	11	1.796	2.201	3.106
12	1.356	1.783	2.681	12	1.783	2.179	3.055
13	1.35	1.771	2.651	13	1.771	2.161	3.013
14	1.345	1.762	2.625	14	1.762	2.145	2.977
15	1.341	1.753	2.603	15	1.753	2.132	2.947
16	1.337	1.746	2.584	16	1.746	2.12	2.921
17	1.334	1.74	2.567	17	1.74	2.11	2.898
18	1.331	1.734	2.553	18	1.734	2.101	2.879
19	1.328	1.729	2.54	19	1.729	2.093	2.861
20	1.326	1.725	2.528	20	1.725	2.086	2.846
21	1.323	1.721	2.518	21	1.721	2.08	2.832
22	1.321	1.717	2.509	22	1.717	2.074	2.819

One-Tailed Test				Two-Tailed Test			
df	0.10	0.05	0.01	df	0.10	0.05	0.01
23	1.32	1.714	2.5	23	1.714	2.069	2.808
24	1.318	1.711	2.492	24	1.711	2.064	2.797
25	1.317	1.708	2.485	25	1.708	2.06	2.788
26	1.315	1.706	2.479	26	1.706	2.056	2.779
27	1.314	1.704	2.473	27	1.704	2.052	2.771
28	1.313	1.701	2.467	28	1.701	2.049	2.764
29	1.312	1.699	2.462	29	1.699	2.045	2.757
30	1.311	1.698	2.458	30	1.698	2.043	2.75
35	1.306	1.69	2.438	35	1.69	2.03	2.724
40	1.303	1.684	2.424	40	1.684	2.021	2.705
45	1.301	1.68	2.412	45	1.68	2.014	2.69
50	1.299	1.676	2.404	50	1.676	2.009	2.678
55	1.297	1.673	2.396	55	1.673	2.004	2.668
60	1.296	1.671	2.39	60	1.671	2.001	2.661
65	1.295	1.669	2.385	65	1.669	1.997	2.654
70	1.294	1.667	2.381	70	1.667	1.995	2.648
75	1.293	1.666	2.377	75	1.666	1.992	2.643
80	1.292	1.664	2.374	80	1.664	1.99	2.639
85	1.292	1.663	2.371	85	1.663	1.989	2.635
90	1.291	1.662	2.369	90	1.662	1.987	2.632
95	1.291	1.661	2.366	95	1.661	1.986	2.629
100	1.29	1.66	2.364	100	1.66	1.984	2.626
Infinity	1.282	1.645	2.327	Infinity	1.645	1.96	2.576

TABLE B.3: CRITICAL VALUES FOR ANALYSIS OF VARIANCE OR F TEST

How to use this table:

1. Compute the F value.

2. Determine the number of degrees of freedom for the numerator $(k - 1)$ and the number of degrees of freedom for the denominator $(n - k)$.

3. Locate the critical value by reading across to locate the degrees of freedom in the numerator and down to locate the degrees of freedom in the denominator. The critical value is at the intersection of this column and row.

4. If the obtained value is greater than the critical or tabled value, the null hypothesis (that the means are equal to one another) is not the most attractive explanation for any observed differences.

5. If the obtained value is less than the critical or tabled value, the null hypothesis is the most attractive explanation for any observed differences.

TABLE B.3   Critical Values for Analysis of Variance or *F* Test

df *for the* Denominator	Type I Error Rate	df *for the Numerator*					
		1	*2*	*3*	*4*	*5*	*6*
1	.01	4052.00	4999.00	5403.00	5625.00	5764.00	5859.00
	.05	162.00	200.00	216.00	225.00	230.00	234.00
	.10	39.90	49.50	53.60	55.80	57.20	58.20
2	.01	98.50	99.00	99.17	99.25	99.30	99.33
	.05	18.51	19.00	19.17	19.25	19.30	19.33
	.10	8.53	9.00	9.16	9.24	9.29	9.33
3	.01	34.12	30.82	29.46	28.71	28.24	27.91
	.05	10.13	9.55	9.28	9.12	9.01	8.94
	.10	5.54	5.46	5.39	5.34	5.31	5.28
4	.01	21.20	18.00	16.70	15.98	15.52	15.21
	.05	7.71	6.95	6.59	6.39	6.26	6.16
	.10	.55	4.33	4.19	4.11	4.05	4.01
5	.01	16.26	13.27	12.06	11.39	10.97	10.67
	.05	6.61	5.79	5.41	5.19	5.05	4.95
	.10	4.06	3.78	3.62	3.52	3.45	3.41
6	.01	13.75	10.93	9.78	9.15	8.75	8.47
	.05	5.99	5.14	4.76	4.53	4.39	4.28
	.10	3.78	3.46	3.29	3.18	3.11	3.06
7	.01	12.25	9.55	8.45	7.85	7.46	7.19
	.05	5.59	4.74	4.35	4.12	3.97	3.87
	.10	3.59	3.26	3.08	2.96	2.88	2.83
8	.01	11.26	8.65	7.59	7.01	6.63	6.37
	.05	5.32	4.46	4.07	3.84	3.69	3.58
	.10	3.46	3.11	2.92	2.81	2.73	2.67
9	.01	10.56	8.02	6.99	6.42	6.06	5.80
	.05	5.12	4.26	3.86	3.63	3.48	3.37
	.10	3.36	3.01	2.81	2.69	2.61	2.55
10	.01	10.05	7.56	6.55	6.00	5.64	5.39
	.05	4.97	4.10	3.71	3.48	3.33	3.22
	.10	3.29	2.93	2.73	2.61	2.52	2.46
11	.01	9.65	7.21	6.22	5.67	5.32	5.07
	.05	4.85	3.98	3.59	3.36	3.20	3.10
	.10	3.23	2.86	2.66	2.54	2.45	2.39
12	.01	9.33	6.93	5.95	5.41	5.07	4.82
	.05	4.75	3.89	3.49	3.26	3.11	3.00
	.10	3.18	2.81	2.61	2.48	2.40	2.33
13	.01	9.07	6.70	5.74	5.21	4.86	4.62
	.05	4.67	3.81	3.41	3.18	3.03	2.92
	.10	3.14	2.76	2.56	2.43	2.35	2.28
14	.01	8.86	6.52	5.56	5.04	4.70	4.46
	.05	4.60	3.74	3.34	3.11	2.96	2.85
	.10	3.10	2.73	2.52	2.40	2.31	2.24
15	.01	8.68	6.36	5.42	4.89	4.56	4.32
	.05	4.54	3.68	3.29	3.06	2.90	2.79
	.10	3.07	2.70	2.49	2.36	2.27	2.21

(Continued)

TABLE B.3 (Continued)

df for the Denominator	Type I Error Rate	df for the Numerator					
		1	2	3	4	5	6
16	.01	8.53	6.23	5.29	4.77	4.44	4.20
	.05	4.49	3.63	3.24	3.01	2.85	2.74
	.10	3.05	2.67	2.46	2.33	2.24	2.18
17	.01	8.40	6.11	5.19	4.67	4.34	4.10
	.05	4.45	3.59	3.20	2.97	2.81	2.70
	.10	3.03	2.65	2.44	2.31	2.22	2.15
18	.01	8.29	6.01	5.09	4.58	4.25	4.02
	.05	4.41	3.56	3.16	2.93	2.77	2.66
	.10	3.01	2.62	2.42	2.29	2.20	2.13
19	.01	8.19	5.93	5.01	4.50	4.17	3.94
	.05	4.38	3.52	3.13	2.90	2.74	2.63
	.10	2.99	2.61	2.40	2.27	2.18	2.11
20	.01	8.10	5.85	4.94	4.43	4.10	3.87
	.05	4.35	3.49	3.10	2.87	2.71	2.60
	.10	2.98	2.59	2.38	2.25	2.16	2.09
21	.01	8.02	5.78	4.88	4.37	4.04	3.81
	.05	4.33	3.47	3.07	2.84	2.69	2.57
	.10	2.96	2.58	2.37	2.23	2.14	2.08
22	.01	7.95	5.72	4.82	4.31	3.99	3.76
	.05	4.30	3.44	3.05	2.82	2.66	2.55
	.10	2.95	2.56	2.35	2.22	2.13	2.06
23	.01	7.88	5.66	4.77	4.26	3.94	3.71
	.05	4.28	3.42	3.03	2.80	2.64	2.53
	.10	2.94	2.55	2.34	2.21	2.12	2.05
24	.01	7.82	5.61	4.72	4.22	3.90	3.67
	.05	4.26	3.40	3.01	2.78	2.62	2.51
	.10	2.93	2.54	2.33	2.20	2.10	2.04
25	.01	7.77	5.57	4.68	4.18	3.86	3.63
	.05	4.24	3.39	2.99	2.76	2.60	2.49
	.10	2.92	2.53	2.32	2.19	2.09	2.03
26	.01	7.72	5.53	4.64	4.14	3.82	3.59
	.05	4.23	3.37	2.98	2.74	2.59	2.48
	.10	2.91	2.52	2.31	2.18	2.08	2.01
27	.01	7.68	5.49	4.60	4.11	3.79	3.56
	.05	4.21	3.36	2.96	2.73	2.57	2.46
	.10	2.90	2.51	2.30	2.17	2.07	2.01
28	.01	7.64	5.45	4.57	4.08	3.75	3.53
	.05	4.20	3.34	2.95	2.72	2.56	2.45
	.10	2.89	2.50	2.29	2.16	2.07	2.00
29	.01	7.60	5.42	4.54	4.05	3.73	3.50
	.05	4.18	3.33	2.94	2.70	2.55	2.43
	.10	2.89	2.50	2.28	2.15	2.06	1.99
30	.01	7.56	5.39	4.51	4.02	3.70	3.47
	.05	4.17	3.32	2.92	2.69	2.53	2.42
	.10	2.88	2.49	2.28	2.14	2.05	1.98

df for the Denominator	Type I Error Rate	df for the Numerator					
		1	2	3	4	5	6
35	.01	7.42	5.27	4.40	3.91	3.59	3.37
	.05	4.12	3.27	2.88	2.64	2.49	2.37
	.10	2.86	2.46	2.25	2.14	2.02	1.95
40	.01	7.32	5.18	4.31	3.91	3.51	3.29
	.05	4.09	3.23	2.84	2.64	2.45	2.34
	.10	2.84	2.44	2.23	2.11	2.00	1.93
45	.01	7.23	5.11	4.25	3.83	3.46	3.23
	.05	4.06	3.21	2.81	2.61	2.42	2.31
	.10	2.82	2.43	2.21	2.09	1.98	1.91
50	.01	7.17	5.06	4.20	3.77	3.41	3.19
	.05	4.04	3.18	2.79	2.58	2.40	2.29
	.10	2.81	2.41	2.20	2.08	1.97	1.90
55	.01	7.12	5.01	4.16	3.72	3.37	3.15
	.05	4.02	3.17	2.77	2.56	2.38	2.27
	.10	2.80	2.40	2.19	2.06	1.96	1.89
60	.01	7.08	4.98	4.13	3.68	3.34	3.12
	.05	4.00	3.15	2.76	2.54	2.37	2.26
	.10	2.79	2.39	2.18	2.05	1.95	1.88
65	.01	7.04	4.95	4.10	3.65	3.31	3.09
	.05	3.99	3.14	2.75	2.53	2.36	2.24
	.10	2.79	2.39	2.17	2.04	1.94	1.87
70	.01	7.01	4.92	4.08	3.62	3.29	3.07
	.05	3.98	3.13	2.74	2.51	2.35	2.23
	.10	2.78	2.38	2.16	2.03	1.93	1.86
75	.01	6.99	4.90	4.06	3.60	3.27	3.05
	.05	3.97	3.12	2.73	2.50	2.34	2.22
	.10	2.77	2.38	2.16	2.03	1.93	1.86
80	.01	3.96	4.88	4.04	3.56	3.26	3.04
	.05	6.96	3.11	2.72	2.49	2.33	2.22
	.10	2.77	2.37	2.15	2.02	1.92	1.85
85	.01	6.94	4.86	4.02	3.55	3.24	3.02
	.05	3.95	3.10	2.71	2.48	2.32	2.21
	.10	2.77	2.37	2.15	2.01	1.92	1.85
90	.01	6.93	4.85	4.02	3.54	3.23	3.01
	.05	3.95	3.10	2.71	2.47	2.32	2.20
	.10	2.76	2.36	2.15	2.01	1.91	1.84
95	.01	6.91	4.84	4.00	3.52	3.22	3.00
	.05	3.94	3.09	2.70	2.47	2.31	2.20
	.10	2.76	2.36	2.14	2.01	1.91	1.84
100	.01	6.90	4.82	3.98	3.51	3.21	2.99
	.05	3.94	3.09	2.70	2.46	2.31	2.19
	.10	2.76	2.36	2.14	2.00	1.91	1.83
Infinity	.01	6.64	4.61	3.78	3.32	3.02	2.80
	.05	3.84	3.00	2.61	2.37	2.22	2.10
	.10	2.71	2.30	2.08	1.95	1.85	1.78

TABLE B.4: VALUES OF THE CORRELATION COEFFICIENT NEEDED FOR REJECTION OF THE NULL HYPOTHESIS

How to use this table:

1. Compute the value of the correlation coefficient.

2. Compare the value of the correlation coefficient with the critical value listed in this table.

3. If the obtained value is greater than the critical or tabled value, the null hypothesis (that the correlation coefficient is equal to 0) is not the most attractive explanation for any observed differences.

4. If the obtained value is less than the critical or tabled value, the null hypothesis is the most attractive explanation for any observed differences.

TABLE B.4   Values of the Correlation Coefficient Needed for Rejection of the Null Hypothesis

	One-Tailed Test			Two-Tailed Test	
df	.05	.01	df	.05	.01
1	.9877	.9995	1	.9969	.9999
2	.9000	.9800	2	.9500	.9900
3	.8054	.9343	3	.8783	.9587
4	.7293	.8822	4	.8114	.9172
5	.6694	.832	5	.7545	.8745
6	.6215	.7887	6	.7067	.8343
7	.5822	.7498	7	.6664	.7977
8	.5494	.7155	8	.6319	.7646
9	.5214	.6851	9	.6021	.7348
10	.4973	.6581	10	.5760	.7079
11	.4762	.6339	11	.5529	.6835
12	.4575	.6120	12	.5324	.6614
13	.4409	.5923	13	.5139	.6411
14	.4259	.5742	14	.4973	.6226
15	.412	.5577	15	.4821	.6055
16	.4000	.5425	16	.4683	.5897
17	.3887	.5285	17	.4555	.5751
18	.3783	.5155	18	.4438	.5614
19	.3687	.5034	19	.4329	.5487
20	.3598	.4921	20	.4227	.5368
25	.3233	.4451	25	.3809	.4869
30	.2960	.4093	30	.3494	.4487
35	.2746	.3810	35	.3246	.4182
40	.2573	.3578	40	.3044	.3932
45	.2428	.3384	45	.2875	.3721
50	.2306	.3218	50	.2732	.3541
60	.2108	.2948	60	.2500	.3248
70	.1954	.2737	70	.2319	.3017
80	.1829	.2565	80	.2172	.2830
90	.1726	.2422	90	.2050	.2673
100	.1638	.2301	100	.1946	.2540

TABLE B.5: CRITICAL VALUES FOR THE CHI-SQUARE TEST

How to use this table:

1. Compute the χ^2 value.

2. Determine the number of degrees of freedom for the rows $(R-1)$ and the number of degrees of freedom for the columns $(C-1)$. If it's a one-dimension table, then you have only columns.

3. Locate the critical value by locating the degrees of freedom in the titled (df) column, and under the appropriate column for level of significance, read across.

4. If the obtained value is greater than the critical or tabled value, the null hypothesis (that the frequencies are equal to one another) is not the most attractive explanation for any observed differences.

5. If the obtained value is less than the critical or tabled value, the null hypothesis is the most attractive explanation for any observed differences.

TABLE B.5   Critical Values for the Chi-Square Test

df	*Level of Significance*		
	.10	*.05*	*.01*
1	2.71	3.84	6.64
2	4.00	5.99	9.21
3	6.25	7.82	11.34
4	7.78	9.49	13.28
5	9.24	11.07	15.09
6	10.64	12.59	16.81
7	12.02	14.07	18.48
8	13.36	15.51	20.09
9	14.68	16.92	21.67
10	16.99	18.31	23.21
11	17.28	19.68	24.72
12	18.65	21.03	26.22
13	19.81	22.36	27.69
14	21.06	23.68	29.14
15	22.31	25.00	30.58
16	23.54	26.30	32.00
17	24.77	27.60	33.41
18	25.99	28.87	34.80
19	27.20	30.14	36.19
20	28.41	31.41	37.57
21	29.62	32.67	38.93
22	30.81	33.92	40.29
23	32.01	35.17	41.64
24	33.20	36.42	42.98
25	34.38	37.65	44.81
26	35.56	38.88	45.64
27	36.74	40.11	46.96
28	37.92	41.34	48.28
29	39.09	42.56	49.59
30	40.26	43.77	50.89

APPENDIX C

Data Sets

Chapter 2 Data Set 1

Score 1	Score 2	Score 3
3	34	154
7	54	167
5	17	132
4	26	145
5	34	154
6	25	145
7	14	113
8	24	156
6	25	154
5	23	123

Chapter 3 Data Set 1

Height	Weight	Height	Weight
53	156	57	154
46	131	68	166
54	123	65	153
44	142	66	140
56	156	54	143
76	171	66	156
87	143	51	173
65	135	58	143
45	138	49	161
44	114	48	131

Chapter 4 Data Set 1

	Comp Score	
12	42	15
15	44	16
11	47	22
16	54	29
21	55	29
25	51	54
21	56	56
8	53	57
6	57	59
2	49	54
22	45	56
26	45	43
27	47	44
36	43	41
34	31	42
33	12	7
38	14	

Chapter 5 Data Set 1

Correct	Attitude
17	94
13	73
12	59
15	80
16	93
14	85
16	66
16	79
18	77
19	91

Chapter 5 Data Set 2

Years of Training	Successful Outcomes
1	9
9	1
1	8
4	7
3	6
3	7
7	9
9	5
7	5
6	6
6	7
1	4

Chapter 6 Data Set 1

Fall Results	Spring Results	Fall Results	Spring Results
21	7	3	30
38	13	16	26
15	35	34	43
34	45	50	20
5	19	14	22
32	47	14	25
24	34	3	50
3	1	4	17
17	12	42	32
32	41	28	46
33	3	40	10
15	20	40	48
21	39	12	11
8	46	5	23

Chapter 10 Data Set 1

Males	Females
9	3
8	5
4	1
9	2
3	6
8	4
10	3
8	6
9	7
8	9
10	7
7	3
6	7
12	6
	8
	8

Chapter 10 Data Set 2

Urban	Rural
6.5	7.9
9.9	4.3
6.8	6.8
4.8	6.5
4	3.3
5.3	13.2
8	9.3
4.2	1.3
7	6.7
6	5.3
9.3	2.4
6.4	4.3
9	1
5.6	3.5
6.6	
5	

Chapter 10 Data Set 3

Sales	Main Street Store	Mall Store
Week 1	$3,453	$2,542
Week 2	$5,435	$3,221
Week 3	$3,656	$1,423
Week 4	$4,543	$1,656
Week 5	$4,543	$4,324
Week 6	$1,232	$3,234
Week 7	$4,543	$2,312
Week 8	$5,643	$1,324
Week 9	$4,354	$2,178
Week 10	$6,342	$5,468
Week 11	$4,355	$2,432
Week 12	$3,232	$2,123
Week 13	$6,532	$1,543
Week 14	$3,234	$1,121
Week 15	$3,545	$4,231

Chapter 11 Data Set 1

Before Recycling	After Recycling	Before Recycling	After Recycling
20	23	23	22
6	8	33	35
12	11	44	41
34	35	65	56
55	57	43	34
43	76	53	51
54	54	22	21
24	26	34	31
33	35	32	33
21	26	44	38
34	28	17	15
33	31	28	27
54	56		

Chapter 11 Data Set 2

Before Intervention	After Intervention	Before Intervention	After Intervention
1.3	6.5	9	8.4
2.5	8.7	7.6	6.4
2.3	9.8	4.5	7.2
8.1	10.2	1.1	5.8
5	7.9	5.6	6.9
7	6.5	6.2	5.9
7.5	8.7	7	7.6
5.2	7.9	6.9	7.8
4.4	8.7	5.6	7.3
7.6	9.1	5.2	4.6

Chapter 12 Data Set 1

<15 Hours of Practice	15–25 Hours of Practice	> 25 Hours of Practice
58.7	64.4	68
55.3	55.8	65.9
61.8	58.7	54.7
49.5	54.7	53.6
64.5	52.7	58.7
61	67.8	58.7
65.7	61.6	65.7
51.4	58.7	66.5
53.6	54.6	56.7
59	51.5	55.4
	54.7	51.5
	61.4	54.8
	56.9	57.2

Chapter 13 Data Set 1

	Treatment 1	Treatment 2
Male	76	88
	78	76
	76	76
	76	76
	76	56
	74	76
	74	76
	76	98
	76	88
	55	78
Female	65	65
	90	67
	65	67
	90	87
	65	78
	90	56
	90	54
	79	56
	70	54
	90	56

Chapter 13 Data Set 2

	Treatment 1	Treatment 2	Treatment 3
Severity 1	6	6	2
	6	5	1
	7	4	3
	7	5	4
	7	4	5
	6	3	4
	5	3	3
	6	3	3
	7	4	3
	8	5	4
	7	5	5
	6	5	3
	5	6	1

	Treatment 1	Treatment 2	Treatment 3
	6	6	2
	7	7	4
	8	6	3
	9	5	5
	8	7	4
	7	6	2
	7	8	3
Severity 2	7	7	4
	8	5	5
	8	4	6
	9	3	5
	8	4	4
	7	5	4
	6	4	6
	6	4	5
	6	3	4
	7	3	2
	7	4	1
	6	5	3
	7	6	2
	8	7	2
	8	7	3
	8	6	4
	9	5	3
	0	4	2
	9	4	2
	8	5	1

Chapter 14 Data Set 1

Motivation	GPA	Motivation	GPA
1	3.4	6	2.6
6	3.4	7	2.5
2	2.5	7	2.8
7	3.1	2	1.8
5	2.8	9	3.7
4	2.6	8	3.1
3	2.1	8	2.5
1	1.6	7	2.4
8	3.1	6	2.1
6	2.6	9	4.0
5	3.2	7	3.9
6	3.1	8	3.1
5	3.2	7	3.3
5	2.7	8	3.0
6	2.8	9	2.0

Chapter 15 Data Set 1

Time	Correct	Time	Correct
14.5	5	13.9	3
13.4	7	17.3	12
12.7	6	12.5	5
16.4	2	16.7	4
21.0	4	22.7	3

APPENDIX D

The Reward

The Brownie Recipe

What the heck is a brownie recipe doing in an introductory statistics book? Good question.

In all seriousness, you have probably worked hard on this material whether for a course, a review, or just working on your own. And, because of all your effort, you deserve a reward, and here it is. The recipe is a version based on several different recipes and some tweaking, and it's all your author's and he is happy to share it with you. There, the secret is out.

These brownies need to age. Right out of the pan, not even cooled, they are terrific on ice cream. When aged a bit, they get very nice and chewy and are great from the fridge. If you freeze them, note that it takes more calories to defrost them in your mouth then are contained in the brownies themselves, so there is a net loss. Eat as many frozen as you want.

1 stick butter

4 ounces unsweetened chocolate (or more)

½ tablespoon of salt

2 eggs

1 cup flour

2 cups sugar

1 tablespoon vanilla

2 tablespoons mayonnaise (I know)

6 ounces chocolate chips (or more)

1 cup whole walnuts

How to do it . . .

1. Preheat oven to 325 °F.

2. Melt unsweetened chocolate and butter in a saucepan.

3. Mix flour and salt together in a bowl.

4. Add sugar, vanilla, nuts, mayonnaise, and eggs to melted chocolate-butter stuff and mix well.

5. Add all of #4 to flour mixture and mix well.

6. Add chocolate chips.

7. Pour into an 8" × 8" greased baking dish.

8. Bake for about 35–40 minutes or until tester comes out clean.

NOTES . . .

- I know about the mayonnaise thing. If you think it sounds weird, then don't put it in. These brownies are not delicious for nothing, though, so leave out this ingredient at your own risk.
- Use good chocolate—the higher the fat content the better, and you can use up to 6 ounces of unsweetened and even more chocolate chips.

GLOSSARY

Analysis of variance

A test for the difference between two or more means. A simple analysis of variance (or ANOVA) has only one independent variable, whereas a factorial analysis of variance tests the means of more than one independent variable. One-way analysis of variance looks for differences between the means of more than two groups.

Arithmetic mean

A measure of central tendency that sums all the scores in the data sets and divides by the number of scores.

Asymptotic

The quality of the normal curve such that the tails never touch.

AVERAGE

A function that returns the mean of its arguments.

Average

The most representative score in a set of scores.

Bell-shaped curve

A distribution of scores that is symmetrical about the mean, median, and mode and has asymptotic tails.

Bivariate correlation

A correlation between two variables.

Cell

The intersection for a row and a column.

Chi-square

A nonparametric test that allows you to determine if what you observe in a distribution of frequencies would be what you would expect to occur by chance.

CHIDIST

A function that returns the one-tailed probability of the chi-squared distribution.

CHITEST

A function that returns the test for independence.

Class interval
The upper and lower boundaries of a set of scores used in the creation of a frequency distribution.

Coefficient of alienation
The amount of variance unaccounted for in the relationship between two variables.

Coefficient of determination
The amount of variance accounted for in the relationship between two variables.

Coefficient of nondetermination
See Coefficient of alienation

Concurrent validity
A type of validity that examines how well a test outcome is consistent with a criterion that occurs in the present.

Construct validity
A type of validity that examines how well a test reflects an underlying construct.

Content validity
A type of validity that examines how well a test samples a universe of items.

CORREL
A function that returns the correlation coefficient between two data sets.

Correlation coefficient
A numerical index that reflects the relationship between two variables.

Correlation matrix
A set of correlation coefficients.

Criterion
Another term for the outcome variable.

Criterion validity
A type of validity that examines how well a test reflects some criterion that occurs either in the present (concurrent) or in the future (predictive).

Critical value
The value necessary for rejection (or nonacceptance) of the null hypothesis.

Cumulative frequency distribution
A frequency distribution that shows frequencies for class intervals along with the cumulative frequency for each.

Data
A record of an observation or an event such as a test score, a grade in math class, or response time.

Data point
An observation.

Data set
A set of data points.

Degrees of freedom
A value that is different for different statistical tests and approximates the sample size of a number of individual cells in an experimental design.

Dependent variable
The outcome variable or the predicted variable in a regression equation.

Descriptive statistics
Values that describe the characteristics of a sample or population.

Direct correlation
A positive correlation where the values of both variables change in the same direction.

Directional research hypothesis
A research hypothesis that includes a statement of inequality.

Effect Size
Effect size is a measure of the magnitude (and not necessarily the size) of the difference between two statistics such as group means.

Error in prediction
The difference between the actual score (Y) and the predicted score (Y').

Error of estimate
See Error in prediction

Error score
The part of a test score that is random and contributes to the unreliability of a test.

Factorial analysis of variance
An analysis of variance with more than one factor or independent variable.

Factorial design
A research design where there is more than one treatment variable.

FDIST
A function that returns the F probability distribution.

FORECAST
A function that returns a value along a linear trend.

Formula
A series of cell references and operators that produces a particular outcome.

Formula bar
   The location on the spreadsheet below the toolbars where cell contents are revealed.

FREQUENCY
   A function that returns a frequency distribution as a vertical array.

Frequency distribution
   A method for illustrating the distribution of scores within class intervals.

Frequency polygon
   A graphical representation of a frequency distribution.

FTEST
   A function that returns the result of an F test.

Function
   A predefined formula.

GEOMEAN
   A function that returns the geometric mean.

Goodness-of-fit test
   See One-sample chi-square

Histogram
   A graphical representation of a frequency distribution.

Hypothesis
   An if–then statement of conjecture that relates variables to one another.

Independent variable
   The treatment variable that is manipulated or the predictor variable in a regression equation.

Indirect correlation
   A negative correlation where the values of variables move in opposite directions.

Inferential statistics
   Tools that are used to infer the results based on a sample to a population.

Interaction effect
   The outcome where the effect of one factor is differentiated across another factor.

INTERCEPT
   A function that computes the location where the regression line crosses the x-axis.

Internal consistency reliability
   A type of reliability that examines the one-dimensional nature of an assessment tool.

Interrater reliability
A type of reliability that examines the consistency between raters.

Interval level of measurement
The level of measurement that stipulates there is an equal interval or distance between various points along some underlying continuum.

KURT
A function that returns the kurtosis of a data set.

Kurtosis
The quality of a distribution such that it is flat or peaked.

Leptokurtic
The quality of a normal curve that defines its peakedness.

Line of best fit
The regression line that best fits the actual scores and minimizes the error in prediction.

Linear correlation
A correlation that is best expressed as a straight line.

LINEST
A function that returns the parameters of a linear trend.

Main effect
In analysis of variance, when a factor or an independent variable has a significant effect upon the outcome variable.

Mean
A type of average where scores are summed and divided by the number of observations.

Mean deviation
The average deviation for all scores from the mean of a distribution.

Measurement
The assignment of values to outcomes following a set of rules.

Measures of central tendency
The mean, median, and mode.

MEDIAN
A function that returns the median of the given numbers.

Median
The point at which 50% of the cases in a distribution fall below and 50% fall above.

Midpoint
The central point in a class interval.

MODE
A function that returns the most common value in a data set.

Mode
The most frequently occurring score in a distribution.

Multiple regression
A statistical technique where several variables are used to predict one.

Nominal level of measurement
The level of measurement that stipulates data are categorical in nature.

Nondirectional research hypothesis
A hypothesis that posits no direction, but a difference.

Nonparametric statistics
Distribution-free statistics.

Normal curve
See Bell-shaped curve

NORMSDIST
A function that returns the standard normal cumulative distribution.

Null hypothesis
A statement of equality between sets of variables, numbers, text, and logical values.

Observed score
The score that is recorded or observed.

Obtained value
The value that results from the application of a statistical test.

Ogive
A visual representation of a cumulative frequency distribution.

One-sample chi-square
A chi-square test that includes only one dimension.

One-tailed test
A directional test.

One-way analysis of variance
See Analysis of variance

Ordinal level of measurement
The level of measurement that stipulates data can be, and are, ranked.

Outliers
Those scores in a distribution that are noticeably more extreme than the majority of scores. Exactly which score is an outlier is usually an arbitrary decision made by the researcher.

Parallel forms reliability
A type of reliability that examines the consistency across different forms of the same test.

Parametric statistics
Statistics used for the inference from a sample to a population.

PEARSON
A function that returns the Pearson product moment correlation coefficient.

Pearson product-moment correlation
See Correlation coefficient

Percentile point
The point at or below where a score appears.

Pivot table
A table where the rows and columns can be rearranged and data extracted.

Platykurtic
The quality of a normal curve that defines its flatness.

Population
All the possible subjects or cases of interest.

Power
A construct that has to do with how well a statistical test can detect and reject a null hypothesis when it is false.

Predictive validity
A type of validity that examines how well a test outcome is consistent with a criterion that occurs in the future.

Predictor variable
The variable that predicts an outcome.

QUARTILE
A function that returns the quartile of a data set.

Range
The highest minus the lowest score, and a gross measure of variability. *Exclusive* range is the highest score minus the lowest score. *Inclusive* range is the highest score minus the lowest score plus 1.

Ratio level of measurement
The level of measurement that stipulates there is an absolute zero to the scale.

Regression equation
The equation that defines the points and the line that are closest to the actual scores.

Regression line
The line drawn based on the values in the regression equation.

Reliability
The quality of a test such that it is consistent.

Research hypothesis
A statement of inequality between two variables.

Sample
A subset of a population.

Sampling error
The difference between sample and population values.

Scales of measurement
Different ways of categorizing measurement outcomes.

Scatterplot, or scattergram
A plot of paired data points.

Significance level
The risk set by the researcher for rejecting a null hypothesis when it is true.

Simple analysis of variance
See Analysis of variance

Single factor
See Analysis of variance

SKEW
A function that returns the skewness of a distribution.

Skew, or skewness
The quality of a distribution that defines the disproportionate frequency of certain scores. A longer right tail than left corresponds to a smaller number of occurrences at the high end of the distribution; this is a *positively* skewed distribution. A shorter right tail than left corresponds to a larger number of occurrences at the high end of the distribution; this is a *negatively* skewed distribution.

SLOPE
A function that returns the slope of the linear regression line.

Source table
A listing of sources of variance in an analysis of variance summary table.

Standard deviation
The average deviation from the mean.

Standard error of estimate
A measure of accuracy in prediction.

Standard score
See *z* score

STANDARDIZE
A function that returns a normalized value.

Statistical significance
See Significance level

Statistics
A set of tools and techniques used to organize, describe, and, interpret information.

STDEV
A function that estimates standard deviation based on a sample.

STDEVP
A function that calculates standard deviation based on the entire population.

STEYX
A function that returns the standard error of the predicted *y* value for each *x* in the regression.

TDIST
A function that returns the *t* distribution.

Test of independence
See Two-sample chi-square

Test statistic value
See Obtained value

Test–retest reliability
A type of reliability that examines consistency over time.

TREND
A function that returns values along a linear trend.

True score
The unobservable part of an observed score that reflects the actual ability or behavior.

TTEST
A function that returns the probability associated with a *t* test.

Two-sample chi-square
A chi-square test that uses two dimensions.

Two-tailed test
A test of a nondirectional hypothesis where the direction of the difference is of little importance.

Type I error
The probability of rejecting a null hypothesis when it is true.

Type II error
The probability of accepting a null hypothesis when it is false.

Unbiased estimate
A conservative estimate of a population parameter.

Validity
The quality of a test such that it measures what it says it does.

VAR
A function that estimates variance based on a sample.

Variability
The amount of spread or dispersion in a set of scores.

Variance
The square of the standard deviation, and another measure of a distribution's spread or dispersion.

VARP
A function that calculates variance based on the entire population.

Workbook
A collection of worksheets.

Worksheet
A single Excel spreadsheet.

Y' or Y prime
The predicted Y value.

z score
A raw score that is adjusted for the mean and standard deviation of the distribution from which the raw score comes.

INDEX

ABOUT THE AUTHOR

Neil J. Salkind (PhD, University of Maryland, 1973) is a Professor Emeritus of Psychology and Research in Education at the University of Kansas in Lawrence, Kansas. He completed postdoctoral training at the Bush Child and Family Policy Institute at the University of North Carolina and has authored and coauthored more than 125 scholarly papers and books. Most recently, he is the author of *Theories of Human Development* (2003) and *Statistics for People Who (Think They) Hate Statistics* (2003). He was the editor of *Child Development Abstracts and Bibliography*, published by the Society for Research in Child Development, from 1988 through 2001, and continues to be active in that organization as a member of various committees.

Supporting researchers for more than 40 years

Research methods have always been at the core of SAGE's publishing program. Founder Sara Miller McCune published SAGE's first methods book, *Public Policy Evaluation*, in 1970. Soon after, she launched the *Quantitative Applications in the Social Sciences* series—affectionately known as the "little green books." Always at the forefront of developing and supporting new approaches in methods, SAGE published early groundbreaking texts and journals in the fields of qualitative methods and evaluation.

Today, more than 40 years and two million little green books later, SAGE continues to push the boundaries with a growing list of more than 1,200 research methods books, journals, and reference works across the social, behavioral, and health sciences. Its imprints—Pine Forge Press, home of innovative textbooks in sociology, and Corwin, publisher of PreK–12 resources for teachers and administrators—broaden SAGE's range of offerings in methods. SAGE further extended its impact in 2008 when it acquired CQ Press and its best-selling and highly respected political science research methods list.

From qualitative, quantitative, and mixed methods to evaluation, SAGE is the essential resource for academics and practitioners looking for the latest methods by leading scholars.

For more information, visit **www.sagepub.com**.